Astronomy

4th Edition

by Stephen P. Maran, PhD

Astronomy For Dummies®, 4th Edition

Published by:
John Wiley & Sons, Inc.
111 River Street
Hoboken, NJ 07030-5774
www.wiley.com

Copyright © 2017 by John Wiley & Sons, Inc., Hoboken, New Jersey

Published simultaneously in Canada

For general information on our other products and services, please contact our Customer Care Department within the U.S. at 877-762-2974, outside the U.S. at 317-572-3993, or fax 317-572-4002. For technical support, please visit https://hub.wiley.com/community/support/dummies.

Wiley publishes in a variety of print and electronic formats and by print-on-demand. Some material included with standard print versions of this book may not be included in e-books or in print-on-demand. If this book refers to media such as a CD or DVD that is not included in the version you purchased, you may download this material at http://booksupport.wiley.com. For more information about Wiley products, visit www.wiley.com.

Library of Congress Control Number: 2017947198

ISBN 978-1-119-37424-4 (pbk); ISBN 978-1-119-37438-1 (ebk); ISBN 978-1-119-37441-1 (ebk)

Manufactured in the United States of America

10 9 8 7 6 5 4 3 2 1

Contents at a Glance

Table of Contents

Introduction

stronomy is the study of the sky, the science of cosmic objects and celestial happenings. It's nothing less than the investigation of the nature of the universe we live in. Astronomers carry out the business of astronomy by using backyard telescopes, huge observatory instruments, radio telescopes that detect celestial radio emissions, and satellites orbiting Earth or positioned in space near Earth or another celestial body, such as the Moon or a planet. Scientists send up telescopes in sounding rockets and on unmanned balloons, some instruments travel far into the solar system aboard deep space probes, and some probes gather samples and return them to Earth.

Astronomy can be a professional or amateur activity. About 25,000 professional astronomers engage in space science worldwide, and an estimated 500,000 amateur astronomers live around the globe. Many of the amateurs belong to local or national astronomy clubs in their home countries.

Professional astronomers conduct research on the Sun and the solar system, the Milky Way galaxy, and the universe beyond. They teach in universities, design satellites in government labs, and operate planetariums. They also write books like this one (but maybe not as good). Most hold PhDs. Nowadays, many professional astronomers study abstruse physics of the cosmos or work with automated, remotely controlled telescopes, so they may not even know the constellations.

Amateur astronomers know the constellations. They share an exciting hobby. Some stargaze on their own; many others join astronomy clubs and organizations of every description. The clubs pass on know-how from old hands to new members, share telescopes and equipment, and hold meetings where members tell about their recent observations or hear lectures by visiting scientists.

Amateur astronomers also hold observing meetings where everyone brings a telescope (or looks through another observer's scope). The amateurs conduct these sessions at regular intervals (such as the first Saturday night of each month) or on special occasions (such as the return of a major meteor shower each August or the appearance of a bright comet like Hale-Bopp). And they save up for really big events, such as a total eclipse of the Sun, when thousands of amateurs and dozens of pros travel across Earth to position themselves in the path of totality and witness one of nature's greatest spectacles.

About This Book

This book explains all you need to know to launch into the great hobby of astronomy. It gives you a leg up on understanding the basic science of the universe as well. The latest space missions will make more sense to you: You'll understand why NASA and other organizations send space probes to planets like Saturn, why robot rovers land on Mars, and why scientists seek samples of the dust in the tail of a comet. You'll know why the Hubble Space Telescope peers out into space and how to check up on other space missions. And when astronomers show up in the newspaper or on television to report their latest discoveries — from space; from the big telescopes in Arizona, Hawaii, Chile, and California; or from radio telescopes in New Mexico, Puerto Rico, Australia, or other observatories around the world — you'll understand the background and appreciate the news. You'll even be able to explain it to your friends.

Read only the parts you want, in any order you want. I explain what you need as you go. Astronomy is fascinating and fun, so keep reading. Before you know it, you'll be pointing out Jupiter, spotting famous constellations and stars, and tracking the International Space Station as it whizzes by overhead. The neighbors may start calling you "stargazer." Police officers may ask you what you're doing in the park at night or why you're standing on the roof with binoculars. Tell 'em you're an astronomer. They probably haven't heard that one (I hope they believe you!).

Foolish Assumptions

You may be reading this book because you want to know what's up in the sky or what the scientists in the space program are doing. Perhaps you've heard that astronomy is a neat hobby, and you want to see whether the rumor is true. Perhaps you want to find out what equipment you need.

You're not a scientist. You just enjoy looking at the night sky and have fallen under its spell, wanting to see and understand the real beauty of the universe.

You want to observe the stars, but you also want to know what you're seeing. Maybe you even want to make a discovery of your own. You don't have to be an astronomer to spot a new comet, and you can even help listen for E.T. Whatever your goal, this book helps you achieve it.

Icons Used in This Book

Throughout this book, helpful icons highlight particularly useful information — even if they just tell you to not sweat the tough stuff. Here's what each symbol means.

REMEMBER

The Remember icon points out information you should file away for future reference.

TECHNICAL STUFF

This nerdy guy appears beside discussions that you can skip if you just want to know the basics and start watching the skies. The scientific background can be good to know, but many people happily enjoy their stargazing without knowing about the physics of supernovas, the mathematics of galaxy chasing, and the ins and outs of dark energy.

TIP

This lightbulb puts you right on track to make use of some inside information as you start skywatching or make progress in the hobby.

WARNING

How much trouble can you get into watching the stars? Not much, if you're careful. But some things you can't be too careful about. This icon alerts you to pay attention so you don't get burned.

Beyond the Book

In addition to the book you're reading right now, be sure to check out the free Cheat Sheet online. It offers a timeline of notable astronomical events and a list of famous female astronomers. To get this Cheat Sheet, simply go to www.dummies.com and enter "Astronomy For Dummies" in the Search box.

If you want to test your astronomy knowledge, check out the practice quizzes online. Each chapter has a corresponding quiz consisting of multiple choice and true/false questions. I've also turned the glossary into flashcards that let you test your knowledge of astronomy terms.

To gain access to the online content, all you have to do is register. Just follow these simple steps:

1. **Find your PIN access code.**

   - **Print book users:** If you purchased a hard copy of this book, turn to the inside front cover to find your PIN.

- **E-book users:** If you purchased this book as an e-book, you can get your PIN by registering your e-book at dummies.com/go/getaccess. Go to this website, find your book and click it, and answer the validation questions to verify your purchase. Then you'll receive an email with your PIN.

2. **Go to Dummies.com and click** Activate Now.

3. **Find your product (*Astronomy For Dummies, 4th Edition*) and then follow the on-screen prompts to activate your PIN.**

Now you're ready to go! You can come back to the program as often as you want — simply log on with the username and password you created during your initial login. No need to enter the access code a second time.

Tip: If you have trouble with your PIN or can't find it, contact Wiley Product Technical Support at 877-762-2974 or go to http://support.wiley.com.

Your registration is good for one year from the day you activate your PIN. After that time frame has passed, you can renew your registration for a fee. The website gives you all the details about how to do so.

Where to Go from Here

You can start anywhere you want. Worried about the fate of the universe? Start off with the Big Bang (see Chapter 16 if you're really interested).

Or you may want to begin with what's in store for you as you pursue your passion for the stars.

Wherever you start, I hope you continue your cosmic exploration and experience the joy, excitement, enlightenment, and enchantment that people have always found in the skies.

1

Getting Started with Astronomy

Discover the basic elements of astronomy, check out a list of constellations, and get a crash course on gravity.

Find out about the resources available to help you check out the night sky, including organizations, facilities, and equipment.

Get an introduction to astronomical and artificial phenomena that sweep across the night sky, such as meteors, comets, and artificial satellites.

Chapter **1**

Seeing the Light: The Art and Science of Astronomy

Step outside on a clear night and look at the sky. If you're a city dweller or live in a cramped suburb, you see dozens, maybe hundreds, of twinkling stars. Depending on the time of the month, you may also see a full Moon and up to five of the eight planets that revolve around the Sun.

A shooting star or "meteor" may appear overhead. What you actually see is the flash of light from a tiny piece of space dust streaking through the upper atmosphere.

Another pinpoint of light moves slowly and steadily across the sky. Is it a space satellite, such as the Hubble Space Telescope, the International Space Station, or just a high-altitude airliner? If you have a pair of binoculars, you may be able to see the difference. Most airliners have running lights, and their shapes may be perceptible.

If you live in the country — on the seashore away from resorts and developments, on the plains, or in the mountains far from any floodlit ski slope — you can see thousands of stars. The Milky Way appears as a beautiful pearly swath across the heavens. What you're seeing is the cumulative glow from millions of faint stars, individually indistinguishable with the naked eye. At a great observation place, such as Cerro Tololo in the Chilean Andes, you can see even more stars. They hang like brilliant lamps in a coal black sky, often not even twinkling, like in van Gogh's *Starry Night* painting.

When you look at the sky, you practice astronomy — you observe the universe that surrounds you and try to make sense of what you see. For thousands of years, everything people knew about the heavens they deduced by simply observing the sky. Almost everything that astronomy deals with

>> Is seen from a distance

>> Is discovered by studying the light that comes to you from objects in space

>> Moves through space under the influence of gravity

This chapter introduces you to these concepts (and more).

Astronomy: The Science of Observation

Astronomy is the study of the sky, the science of cosmic objects and celestial happenings, and the investigation of the nature of the universe you live in. Professional astronomers carry out the business of astronomy by observing with telescopes that capture visible light from the stars or by tuning in to radio waves that come from space. They use backyard telescopes, huge observatory instruments, and satellites that orbit Earth collecting forms of light (such as ultraviolet radiation) that the atmosphere blocks from reaching the ground. They send up telescopes in sounding rockets (equipped with instruments for making high-altitude scientific observations) and on unmanned balloons. And they send some instruments into the solar system aboard deep-space probes.

Professional astronomers study the Sun and the solar system, the Milky Way, and the universe beyond. They teach in universities, design satellites in government labs, and operate planetariums. They also write books (like me, your loyal *For Dummies* hero). Most have completed years of schooling to hold PhDs. Many of them study complex physics or work with automated, robotic telescopes that reach far beyond the night sky recognizable to our eyes. They may never have studied the *constellations* (groups of stars, such as Ursa Major, the Great Bear, named by ancient stargazers) that amateur or hobbyist astronomers first explore.

You may already be familiar with the Big Dipper, an asterism in Ursa Major. An *asterism* is a named star pattern that's not identical to one of the 88 recognized constellations. An asterism may be wholly within a single constellation or may include stars from more than one constellation. For example, the four corners of the Great Square of Pegasus, a large asterism, are marked by three stars of the Pegasus constellation and a fourth from Andromeda. Figure 1-1 shows the Big Dipper in the night sky. (In the United Kingdom, some people call the Big Dipper the *Plough*.)

FIGURE 1-1:
The Big Dipper, found in Ursa Major, is an asterism.

In addition to the roughly 30,000 professional astronomers worldwide, several hundred thousand amateur astronomers enjoy watching the skies. Amateur astronomers usually know the constellations and use them as guideposts when exploring the sky by eye, with binoculars, and with telescopes. Many amateurs also make useful scientific contributions. They monitor the changing brightness of variable stars; discover asteroids, comets, and exploding stars; and crisscross Earth to catch the shadows cast as asteroids pass in front of bright stars (thereby helping astronomers map the asteroids' shapes). They even join in professional research efforts with their home computers and smartphones through Citizen Science projects, which I describe in Chapter 2 and elsewhere throughout the book.

In the rest of Part 1, I provide you with information on how to observe the skies effectively and enjoyably.

What You See: The Language of Light

Light brings us information about the planets, moons, and comets in our solar system; the stars, star clusters, and nebulae in our galaxy; and the objects beyond.

In ancient times, folks didn't think about the physics and chemistry of the stars; they absorbed and passed down folk tales and myths: the Great Bear, the Demon star, the Man in the Moon, the dragon eating the Sun during a solar eclipse, and more. The tales varied from culture to culture. But many people did discover the patterns of the stars. In Polynesia, skilled navigators rowed across hundreds of miles of open ocean with no landmarks in view and no compass. They sailed by the stars, the Sun, and their knowledge of prevailing winds and currents.

Gazing at the light from a star, the ancients noted its brightness, position in the sky, and color. This information helps people distinguish one sky object from another, and the ancients (and now people today) got to know them like old friends. Some basics of recognizing and describing what you see in the sky are

>> Distinguishing stars from planets

>> Identifying constellations, individual stars, and other sky objects by name

>> Observing brightness (given as magnitudes)

>> Understanding the concept of a light-year

>> Charting sky position (measured in special units called *RA* and *Dec*)

They wondered as they wandered: Understanding planets versus stars

The term *planet* comes from the ancient Greek word *planetes*, meaning "wanderer." The Greeks (and other ancient people) noticed that five spots of light moved across the pattern of stars in the sky. Some moved steadily ahead; others occasionally looped back on their own paths. Nobody knew why. And these spots of light didn't twinkle like the stars did; no one understood that difference, either. Every culture had a name for those five spots of light — what we now call planets. Their English names are Mercury, Venus, Mars, Jupiter, and Saturn. These celestial bodies aren't wandering through the stars; they orbit around the Sun, our solar system's central star.

Today astronomers know that planets can be smaller or bigger than Earth, but they all are much smaller than the Sun. The planets in our solar system are so close to Earth that they have perceptible disks — at least, when viewed through a telescope — so we can see their shapes and sizes. The stars are so far away from Earth that even if you view them through a powerful telescope, they show up only as points of light. (For more about the planets in the solar system, flip to Part 2. I cover the planets of stars beyond the Sun in Part 4.)

If you see a Great Bear, start worrying: Naming stars and constellations

I used to tell planetarium audiences who craned their necks to look at stars projected above them, "If you can't see a Great Bear up there, don't worry. Maybe those who *do* see a Great Bear should worry."

Ancient astronomers divided the sky into imaginary figures, such as Ursa Major (Latin for "Great Bear"); Cygnus, the Swan; Andromeda, the Chained Lady; and Perseus, the Hero. The ancients identified each figure with a pattern of stars.

The truth is, to most people, Andromeda doesn't look much like a chained lady at all — or anything else, for that matter (see Figure 1-2).

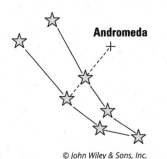

Andromeda

FIGURE 1-2:
Andromeda is also known as the Chained Lady.

Today astronomers have divided the sky into 88 constellations, which contain all the stars you can see. The International Astronomical Union, which governs the science, set boundaries for the constellations so astronomers can agree on which star is in which constellation. Previously, sky maps drawn by different astronomers often disagreed. Now when you read that the Tarantula Nebula is in Dorado (see Chapter 12), you know that, to see this nebula, you must seek it in the Southern Hemisphere constellation Dorado, the Goldfish.

The largest constellation is Hydra, the Water Snake. The smallest is Crux, the Cross, which most people call the Southern Cross. You can see a Northern Cross, too, but you can't find it in a list of constellations; it's an asterism within Cygnus, the Swan. Although astronomers generally agree on the names of the constellations, they don't have a consensus on what each name means. For example, some astronomers call Dorado the Swordfish, but I'd like to skewer that name. One constellation, Serpens, the Serpent, is broken into two sections that aren't connected. The two sections, located on either side of Ophiuchus, the Serpent Bearer, are Serpens Caput (the Serpent's Head) and Serpens Cauda (the Serpent's Tail).

The individual stars in a constellation often have no relation to each other except for their proximity in the sky as visible from Earth. In space, the stars that make up a constellation may be completely unrelated to one another, with some located relatively near Earth and others located at much greater distances in space. But they make a simple pattern for observers on Earth to enjoy.

As a rule, the brighter stars in a constellation were assigned a Greek letter, either by the ancient Greeks or by astronomers of later civilizations. In each constellation, the brightest star was labeled alpha, the first letter of the Greek alphabet. The next brightest star was beta, the second Greek letter, and so on down to omega, the final letter of the 24-character Greek alphabet. (The astronomers used only lowercase Greek letters, so you see them written as α, β, . . . ω.)

So Sirius, the brightest star in the night sky — in Canis Major, the Great Dog — is called Alpha Canis Majoris. (Astronomers add a suffix here or there to put star names in the Latin genitive case; scientists have always liked Latin.) Table 1-1 shows a list of the Greek alphabet, in order, with the names of the letters and their corresponding symbols.

TABLE 1-1

The Greek Alphabet

Letter	Name
α	Alpha
β	Beta
γ	Gamma
δ	Delta
ε	Epsilon
ζ	Zeta
η	Eta
θ	Theta
ι	Iota
κ	Kappa
λ	Lambda
μ	Mu
ν	Nu
ξ	Xi
o	Omicron
π	Pi
ρ	Rho
σ	Sigma
τ	Tau
υ	Upsilon
φ	Phi
χ	Chi
ψ	Psi
ω	Omega

When you look at a star atlas, you discover that the individual stars in a constellation aren't marked α Canis Majoris, β Canis Majoris, and so on. Usually, the creator of the atlas marks the area of the whole constellation as Canis Major and labels the individual stars α, β, and so on. When you read about a star in a list of objects to observe, say, in an astronomy magazine (see Chapter 2), you probably won't see it listed in the style of Alpha Canis Majoris or even α Canis Majoris. Instead, to save space, the magazine prints it as α CMa; *CMa* is the three-letter abbreviation for Canis Majoris (and also the abbreviation for Canis Major). I give the abbreviation for each of the constellations in Table 1-2.

Astronomers didn't coin special names such as Sirius for every star in Canis Major, so they named them with Greek letters or other symbols. In fact, some constellations don't have a single named star. (Don't fall for those advertisements that offer to name a star for a fee. The International Astronomical Union doesn't recognize purchased star names.) In other constellations, astronomers assigned Greek letters, but they could see more stars than the 24 Greek letters. Therefore, astronomers gave some stars Arabic numbers or letters from the Roman alphabet, or numbers in professional catalogues. So you see star names such as 61 Cygni, b Vulpeculae, HR 1516, and more. You may even run across the star names RU Lupi and YY Sex. (I'm not making this up.) But as with any other star, you can recognize them by their positions in the sky (as tabulated in star lists), their brightness, their color, or other properties, if not their names.

When you look at the constellations today, you see many exceptions to the rule that the Greek-letter star names correspond to the respective brightness of the stars in a constellation. The exceptions exist because

>> The letter names were based on inaccurate naked-eye observations of brightness.

>> Over the years, star atlas authors changed constellation boundaries, moving some stars from one constellation into another that included previously named stars.

>> Some astronomers mapped out small and Southern Hemisphere constellations long after the Greek period, and they didn't always follow the lettering practice.

>> The brightness of some stars has changed over the centuries since the ancient Greeks charted them.

A good (or bad) example is the constellation Vulpecula, the Fox, in which only one of the stars (alpha) has a Greek letter.

Because alpha isn't always the brightest star in a constellation, astronomers needed another term to describe that exalted status, and *lucida* is the word (from the Latin word *lucidus*, meaning "bright" or "shining"). The lucida of Canis Major is Sirius, the alpha star, but the lucida of Orion, the Hunter, is Rigel, which is Beta Orionis. The lucida of Leo Minor, the Little Lion (a particularly inconspicuous constellation), is 46 Leo Minoris.

Table 1-2 lists the 88 constellations, the brightest star in each, and the magnitude of that star. *Magnitude* is a measure of a star's brightness. (I talk about magnitudes in the later section "The smaller, the brighter: Getting to the root of magnitudes.") When the lucida of a constellation is the alpha star and has a name, I list only the name. For example, in Auriga, the Charioteer, the brightest star (Alpha Aurigae) is Capella. But when the lucida isn't an alpha, I give its Greek letter or other designation in parentheses. For example, the lucida of Cancer, the Crab, is Al Tarf, which is Beta Cancri.

TIP

If you're a long-time *Astronomy For Dummies* reader (possessing at least one of the three previous editions of the book as well as this edition), you may notice some changes in Table 1-2. In 2016, the International Astronomical Union issued a list of official names for bright stars. Seven stars in Table 1-2 were affected, with minor changes in spelling or a whole new name. In one case, a star was named after its constellation: Alpha Pavonis, in Pavo the Peacock, was itself named Peacock.

TABLE 1-2   ## The Constellations and Their Brightest Stars

Name	Abbreviation	Meaning	Star	Magnitude
Andromeda	And	Chained Lady	Alpheratz	2.1
Antlia	Ant	Air Pump	Alpha Antliae	4.3
Apus	Aps	Bird of Paradise	Alpha Apodis	3.8
Aquarius	Aqr	Water Bearer	Sadalsuud (Beta Aquarii)	2.9
Aquila	Aql	Eagle	Altair	0.8
Ara	Ara	Altar	Beta Arae	2.9
Aries	Ari	Ram	Hamal	2.0
Auriga	Aur	Charioteer	Capella	0.1
Bootes	Boo	Herdsman	Arcturus	−0.04

Name	Abbreviation	Meaning	Star	Magnitude
Caelum	Cae	Chisel	Alpha Caeli	4.5
Camelopardalis	Cam	Giraffe	Beta Camelopardalis	4.0
Cancer	Cnc	Crab	Al Tarf (Beta Cancri)	3.5
Canes Venatici	CVn	Hunting Dogs	Cor Caroli	2.9
Canis Major	CMa	Great Dog	Sirius	−1.5
Canis Minor	CMi	Little Dog	Procyon	0.4
Capricornus	Cap	Goat	Deneb Algedi (Delta Capricorni)	2.9
Carina	Car	Ship's Keel	Canopus	−0.7
Cassiopeia	Cas	Queen	Schedar	2.2
Centaurus	Cen	Centaur	Rigil Kentaurus	−0.01
Cepheus	Cep	King	Alderamin	2.4
Cetus	Cet	Whale	Diphda (Beta Ceti)	2.0
Chamaeleon	Cha	Chameleon	Alpha Chamaeleontis	4.1
Circinus	Cir	Compasses	Alpha Circini	3.2
Columba	Col	Dove	Phact	2.6
Coma Berenices	Com	Berenice's Hair	Beta Comae Berenices	4.3
Corona Australis	CrA	Southern Crown	Alphecca Meridiana	4.1
Corona Borealis	CrB	Northern Crown	Alphecca	2.2
Corvus	Crv	Crow	Gienah (Gamma Corvi)	2.6
Crater	Crt	Cup	Delta Crateris	3.6
Crux	Cru	Cross	Acrux	1.3
Cygnus	Cyg	Swan	Deneb	1.3
Delphinus	Del	Dolphin	Rotanev (Beta Delphini)	3.6
Dorado	Dor	Goldfish	Alpha Doradus	3.3
Draco	Dra	Dragon	Eltanin (Gamma Draconis)	2.2
Equuleus	Equ	Little Horse	Kitalpha	3.9
Eridanus	Eri	River	Achernar	0.5

(continued)

TABLE 1-2 *(continued)*

Name	Abbreviation	Meaning	Star	Magnitude
Fornax	For	Furnace	Alpha Fornacis	3.9
Gemini	Gem	Twins	Pollux (Beta Geminorum)	1.1
Grus	Gru	Crane	Alnair	1.7
Hercules	Her	Hercules	Kornephoros (Beta Herculis)	2.8
Horologium	Hor	Clock	Alpha Horologii	3.9
Hydra	Hya	Water Snake	Alphard	2.0
Hydrus	Hyi	Little Water Snake	Beta Hydri	2.8
Indus	Ind	Indian	Alpha Indi	3.1
Lacerta	Lac	Lizard	Alpha Lacertae	3.8
Leo	Leo	Lion	Regulus	1.4
Leo Minor	LMi	Little Lion	Praecipua (46 Leonis Minoris)	3.8
Lepus	Lep	Hare	Arneb	2.6
Libra	Lib	Scales	Zubeneschamali (Beta Librae)	2.6
Lupus	Lup	Wolf	Alpha Lupi	2.3
Lynx	Lyn	Lynx	Alpha Lyncis	3.1
Lyra	Lyr	Lyre	Vega	0.0
Mensa	Men	Table	Alpha Mensae	5.1
Microscopium	Mic	Microscope	Gamma Microscopii	4.7
Monoceros	Mon	Unicorn	Beta Monocerotis	3.7
Musca	Mus	Fly	Alpha Muscae	2.7
Norma	Nor	Level and Square	Gamma Normae	4.0
Octans	Oct	Octant	Nu Octantis	3.8
Ophiuchus	Oph	Serpent Bearer	Rasalhague	2.1
Orion	Ori	Hunter	Rigel (Beta Orionis)	0.1
Pavo	Pav	Peacock	Peacock	1.9
Pegasus	Peg	Winged Horse	Enif (Epsilon Pegasi)	2.4
Perseus	Per	Hero	Mirfak	1.8

Name	Abbreviation	Meaning	Star	Magnitude
Phoenix	Phe	Phoenix	Ankaa	2.4
Pictor	Pic	Easel	Alpha Pictoris	3.2
Pisces	Psc	Fish	Kullat Nunu (Eta Piscium)	3.6
Pisces Austrinus	PsA	Southern Fish	Fomalhaut	1.2
Puppis	Pup	Ship's Stern	Naos (Zeta Puppis)	2.3
Pyxis	Pyx	Compass	Alpha Pyxidis	3.7
Reticulum	Ret	Net	Alpha Reticuli	3.4
Sagitta	Sge	Arrow	Gamma Sagittae	3.5
Sagittarius	Sgr	Archer	Kaus Australis (Epsilon Sagittarii)	1.9
Scorpius	Sco	Scorpion	Antares	1.0
Sculptor	Scl	Sculptor	Alpha Sculptoris	4.3
Scutum	Sct	Shield	Alpha Scuti	3.9
Serpens	Ser	Serpent	Unukalhai	2.7
Sextans	Sex	Sextant	Alpha Sextantis	4.5
Taurus	Tau	Bull	Aldebaran	0.9
Telescopium	Tel	Telescope	Alpha Telescopii	3.5
Triangulum	Tri	Triangle	Beta Trianguli	3.0
Triangulum Australe	TrA	Southern Triangle	Atria	1.9
Tucana	Tuc	Toucan	Alpha Tucanae	2.9
Ursa Major	UMa	Great Bear	Alioth (Epsilon Ursae Majoris)	1.8
Ursa Minor	UMi	Little Bear	Polaris	2.0
Vela	Vel	Sails	Suhail al Muhlif (Gamma Velorum)	1.8
Virgo	Vir	Virgin	Spica	1.0
Volans	Vol	Flying Fish	Gamma Volantis	3.8
Vulpecula	Vul	Fox	Anser	4.4

Identifying stars would be much easier if they had little name tags that you could see through your telescope. If you have a smartphone, you can download an app to identify the stars for you. Just download a sky map or planetarium app (such as Sky Safari, Star Walk, or Google Sky Map) and face the phone toward the sky. The app generates a map of the constellations in the general direction your phone is facing. With some apps, when you touch the image of a star, its name appears. (I describe more astronomy apps in Chapter 2; for the full scoop on stars, check out Chapter 11.)

What do I spy? Spotting the Messier Catalog and other sky objects

Naming stars was easy enough for astronomers. But what about all those other objects in the sky — galaxies, nebulae, star clusters, and the like (which I cover in Part 3)? Charles Messier (1730–1817), a French astronomer, created a numbered list of about 100 fuzzy sky objects. His list is known as the *Messier Catalog,* and now when you hear the Andromeda Galaxy called by its scientific name, M31, you know that it stands for number 31 in the Catalog. Today 110 objects make up the standard Messier Catalog.

TIP

You can find pictures and a complete list of the Messier objects at The Messier Catalog website of Students for the Exploration and Development of Space at messier.seds.org. And you can find out how to earn a certificate for viewing Messier objects from the Astronomical League Messier Program website at www.astroleague.org/al/obsclubs/messier/mess.html.

Experienced amateur astronomers often engage in Messier marathons, in which each person tries to observe every object in the *Messier Catalog* during a single long night. But in a marathon, you don't have time to enjoy an individual nebula, star cluster, or galaxy. My advice is to take it slow and savor their individual visual delights. A wonderful book on the Messier objects, which includes hints on how to observe each object, is Stephen James O'Meara's *Deep-Sky Companions: The Messier Objects,* 2nd Edition (Cambridge University Press).

Since Messier's time, astronomers have confirmed the existence of thousands of other *deep sky objects,* the term amateurs use for star clusters, nebulae, and galaxies to distinguish them from stars and planets. Because Messier didn't list them, astronomers refer to these objects by their numbers as given in other catalogues. You can find many of these objects listed in viewing guides and sky maps by their NGC *(New General Catalogue)* and IC *(Index Catalogue)* numbers. For example, the bright double cluster in Perseus, the Hero, consists of NGC 869 and NGC 884.

The smaller, the brighter: Getting to the root of magnitudes

A star map, constellation drawing, or list of stars always indicates each star's magnitude. The *magnitudes* represent the brightness of the stars. One of the ancient Greeks, Hipparchos (also spelled Hipparchus, but he wrote it in Greek), divided all the stars he could see into six classes. He called the brightest stars magnitude 1 or *1st magnitude*, the next brightest bunch the *2nd magnitude* stars, and on down to the dimmest ones, which were *6th magnitude.*

Notice that, contrary to most common measurement scales and units, the brighter the star, the smaller the magnitude. The Greeks weren't perfect, however; even Hipparchos had an Achilles' heel: He didn't leave room in his system for the very brightest stars, when accurately measured.

So today we recognize a few stars with a zero magnitude or a negative magnitude. Sirius, for example, is magnitude −1.5. And the brightest planet, Venus, is sometimes magnitude −4 (the exact value differs, depending on the distance Venus is from Earth at the time and its direction with respect to the Sun).

Another omission: Hipparchos didn't have a magnitude class for stars that were too dim to be seen with the naked eye. This didn't seem like an oversight at the time because nobody knew about these stars before the invention of the telescope. But today astronomers know that billions of stars exist beyond our naked-eye view. Their magnitudes are larger numbers: 7 or 8 for stars easily seen through binoculars, and 10 or 11 for stars easily seen through a good, small telescope. The magnitudes reach as high (and as dim) as 21 for the faintest stars in the Palomar Observatory Sky Survey and about 31 for the faintest objects imaged with the Hubble Space Telescope.

Looking back on light-years

The distances to the stars and other objects beyond the planets of our solar system are measured in *light-years.* As a measurement of actual length, a light-year is about 5.9 trillion miles long.

People confuse a light-year with a length of time because the term contains the word *year.* But a light-year is really a distance measurement — the length that light travels, zipping through space at 186,000 miles per second, over the course of a year.

BY THE NUMBERS: THE MATHEMATICS OF BRIGHTNESS

The 1st magnitude stars are about 100 times brighter than the 6th magnitude stars. In particular, the 1st magnitude stars are about 2.512 times brighter than the 2nd magnitude stars, which are about 2.512 times brighter than the 3rd magnitude stars, and so on. (At the 6th magnitude, you get up into some big numbers: 1st magnitude stars are about 100 times brighter.) You mathematicians out there recognize this as a *geometric progression.* Each magnitude is the 5th root of 100 (meaning that when you multiply a number by itself four times — for example, $2.512 \times 2.512 \times 2.512 \times 2.512 \times 2.512$ — the result is 100). If you doubt my word and do this calculation on your own, you get a slightly different answer because I left off some decimal places.

Thus, you can calculate how faint a star is — compared to some other star — from its magnitude. If two stars are 5 magnitudes apart (such as the 1st magnitude star and the 6th magnitude star), they differ by a factor of 2.512^5 (2.512 to the fifth power), and a good pocket calculator shows you that one star is 100 times brighter. If two stars are 6 magnitudes apart, one is about 250 times brighter than the other. And if you want to compare, say, a 1st magnitude star with an 11th magnitude star, you compute a 2.512^{10} difference in brightness, meaning a factor of 100^2, or 10,000.

The faintest object visible with the Hubble Space Telescope is about 25 magnitudes fainter than the faintest star you can see with the naked eye (assuming normal vision and viewing skills — some experts and a certain number of liars and braggarts say that they can see 7th magnitude stars). Speaking of dim stars, 25 magnitudes are five times 5 magnitudes, which corresponds to a brightness difference of a factor of 100^5. So the Hubble can see $100 \times 100 \times 100 \times 100 \times 100$, or 10 billion times fainter than the human eye. Astronomers expect nothing less from a billion-dollar telescope. At least it didn't cost $10 billion.

You can get a good telescope for well under $1,000, and you can view the billion-dollar Hubble's best photos on the Internet for free at hubblesite.org.

When you view an object in space, you see it as it appeared when the light left the object. Consider these examples:

» When astronomers spot an explosion on the Sun, we don't see it in real time; the light from the explosion takes about 8 minutes to get to Earth.

» The nearest star beyond the Sun, Proxima Centauri, is about 4 light-years away. Astronomers can't see Proxima as it is now — only as it was four years ago.

>> Look up at the Andromeda Galaxy, the most distant object that you can readily see with the unaided eye, on a clear, dark night in the fall. The light your eye receives left that galaxy about 2.5 million years ago. If there was a big change in Andromeda tomorrow, we wouldn't know that it happened for more than 2 million years. (See Chapter 12 for hints on viewing the Andromeda Galaxy and other prominent galaxies.)

Here's the bottom line:

>> When you look out into space, you're looking back in time.

>> Astronomers don't have a way to know exactly what an object out in space looks like right now.

When you look at some big, bright stars in a faraway galaxy, you must entertain the possibility that those particular stars don't even exist anymore. As I explain in Chapter 11, some massive stars live for only 10 million or 20 million years. If you see them in a galaxy that is 50 million light-years away, you're looking at lame duck stars. They aren't shining in that galaxy anymore; they're dead.

If astronomers send a flash of light toward one of the most distant galaxies found with Hubble and other major telescopes, the light would take billions of years to arrive. Astronomers, however, calculate that the Sun will swell up and destroy all life on Earth a mere 5 billion or 6 billion years from now, so the light would be a futile advertisement of our civilization's existence, a flash in the celestial pan.

HEY, YOU! NO, NO, I MEAN AU

Earth is about 93 million miles from the Sun, or 1 *astronomical unit* (AU). The distances between objects in the solar system are usually given in AU. Its plural is also AU. (Don't confuse AU with "Hey, you!")

In public announcements, press releases, and popular books, astronomers state how far the stars and galaxies that they study are "from Earth." But among themselves and in technical journals, they always give the distances from the Sun, the center of our solar system. This discrepancy rarely matters because astronomers can't measure the distances of the stars precisely enough for 1 AU more or less to make a difference, but they do it this way for consistency.

Keep on moving: Figuring the positions of the stars

Astronomers used to call stars "fixed stars," to distinguish them from the wandering planets. But in fact, stars are in constant motion as well, both real and apparent. The whole sky rotates overhead because Earth is turning. The stars rise and set, like the Sun and the Moon, but they stay in formation. The stars that make up the Great Bear don't swing over to the Little Dog or Aquarius, the Water Bearer. Different constellations rise at different times and on different dates, as seen from different places around the globe.

Actually, the stars in Ursa Major (and every other constellation) do move with respect to one another — and at breathtaking speeds, measured in hundreds of miles per second. But those stars are so far away that scientists need precise measurements over considerable intervals of time to detect their motions across the sky. So 20,000 years from now, the stars in Ursa Major will form a different pattern in the sky. (Maybe they will even look like a Great Bear.)

In the meantime, astronomers have measured the positions of millions of stars, and many of them are tabulated in catalogs and marked on star maps. The positions are listed in a system called right ascension and declination — known to all astronomers, amateur and pro, as *RA* and *Dec*:

>> The RA is the position of a star measured in the east–west direction on the sky (like longitude, the position of a place on Earth measured east or west of the prime meridian at Greenwich, England).

>> The Dec is the position of the star measured in the north–south direction, like the latitude of a city, which is measured north or south of the equator.

Astronomers usually list RA in units of hours, minutes, and seconds, like time. We list Dec in degrees, minutes, and seconds of arc. Ninety degrees make up a right angle, 60 minutes of arc make up a degree, and 60 seconds of arc equal a minute of arc. A minute or second of arc is also often called an "arc minute" or an "arc second," respectively.

A few simple rules may help you remember how RA and Dec work and how to read a star map (see Figure 1-3):

>> The North Celestial Pole (NCP) is the place to which the axis of Earth points in the north direction. If you stand at the geographic North Pole, the NCP is right overhead. (If you stand there, say "Hi" to Santa for me, but beware: You may be on thin ice because there's no land at the geographic North Pole.)

DIGGING DEEPER INTO RA AND DEC

A star at RA 2h00m00s is 2 hours east of a star at RA 0h00m00s, regardless of their declinations. RA increases from west to east, starting from RA 0h00m00s, which corresponds to a line in the sky (actually half a circle, centered on the center of Earth) from the North Celestial Pole to the South Celestial Pole. The first star may be at Dec 30° North, and the second star may be at Dec 15° 25'12" South, but they're still 2 hours apart in the east–west direction (and 45° 25'12" apart in the north–south direction). The North and South Celestial Poles are the points in the sky — due north and due south — around which the whole sky seems to turn, with the stars all rising and setting.

Note the following details about the units of RA and Dec:

- An hour of RA equals an arc of 15 degrees on the equator in the sky. Twenty-four hours of RA span the sky, and $24 \times 15 = 360$ degrees, or a complete circle around the sky. A minute of RA, called a *minute of time,* is a measure of angle on the sky that makes up $\frac{1}{60}$ of an hour of RA. So you take $15° \div 60$, or $\frac{1}{4}°$. A second of RA, or a *second of time,* is 60 times smaller than a minute of time.

- Dec is measured in degrees, like the degrees in a circle, and in minutes and seconds of arc. A whole degree is about twice the apparent or angular size of the full Moon. Each degree is divided into 60 minutes of arc. The Sun and the full Moon are both about 32 minutes of arc (32') wide, as seen on the sky, although, in reality, the Sun is much larger than the Moon. Each minute of arc is divided into 60 seconds of arc (60"). When you look through a backyard telescope at high magnification, turbulence in the air blurs the image of a star. Under good conditions (low turbulence), the image should measure about 1" or 2" across. That's 1 or 2 arc seconds, not 1 or 2 inches.

» The South Celestial Pole (SCP) is the place to which the axis of Earth points in the south direction. If you stand at the geographic South Pole, the SCP is right overhead. I hope you dressed warmly: You're in Antarctica!

» The imaginary lines of equal RA run through the NCP and SCP as semicircles centered on the center of Earth. They may be imaginary, but they appear marked on most sky maps to help people find the stars at particular RAs.

» The imaginary lines of equal Dec, like the line in the sky that marks Dec of 30° North, pass overhead at the corresponding geographic latitudes. So if you stand in New York City, latitude 41° North, the point overhead is always at Dec 41° North, although its RA changes constantly as Earth turns. These imaginary lines appear on star maps, too, as *declination circles.*

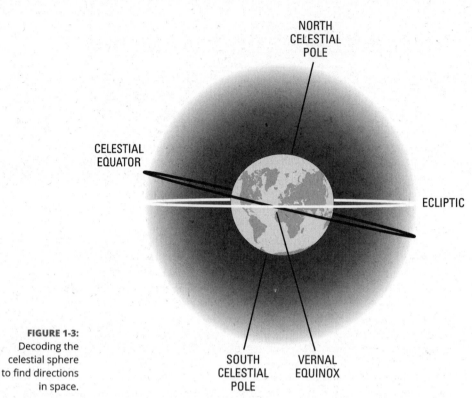

NORTH CELESTIAL POLE

CELESTIAL EQUATOR

ECLIPTIC

SOUTH CELESTIAL POLE

VERNAL EQUINOX

FIGURE 1-3:
Decoding the celestial sphere to find directions in space.

Suppose you want to find the NCP as visible from your backyard. Face due north and look at an altitude of x degrees, where x is your geographic latitude. I'm assuming that you live in North America, Europe, or somewhere in the Northern Hemisphere. If you live in the Southern Hemisphere, you can't see the NCP. You can, however, look for the SCP. Look for the spot due south whose altitude in the sky, measured in degrees above the horizon, is equal to your geographic latitude.

In almost every astronomy book, the symbol " means seconds of arc, not inches. But at every university, a student in Astronomy 101 writes on a lab report, "The image of the star was about 1 inch in diameter." Understanding beats memorizing every day, but not everyone understands.

Here's the good news: If you just want to spot the constellations and the planets, you don't have to know how to use RA and Dec. Just consult a star map drawn for the current week or month (you can find these on the website of *Sky & Telescope* or one of the other magazines that I mention in Chapter 2, in the magazines them-selves, or using a desktop planetarium program for your home computer or a planetarium app for your smartphone or tablet; I recommend programs, websites, and apps in Chapter 2 as well). But if you want to understand how star catalogs and maps work and how to zero in on faint galaxies with your telescope, understanding the system helps.

And if you purchase one of those snazzy and surprisingly affordable telescopes with computer control (see Chapter 3), you can punch in the RA and Dec of a recently discovered comet, and the scope points right at it. (A little table called an *ephemeris* comes with every announcement of a new comet. It gives the predicted RA and Dec of the comet on successive nights as it sweeps across the sky.)

Gravity: A Force to Be Reckoned With

Ever since the work of Sir Isaac Newton, the English scientist (1642–1727), everything in astronomy has revolved around gravity. Newton explained gravity as a force between any two objects. The force depends on mass and separation. The more massive the object, the more powerful its pull. The greater the distance, the weaker the gravitational attraction. Newton sure was a smart cookie!

Albert Einstein developed an improved theory of gravity, which passes experimental tests that Newton's theory flunks. Newton's theory was good enough for commonly experienced gravity, like the force that made the apple fall on his head (if it really hit him). But in other respects, Newton's theory was hit or miss. Einstein's theory is better because it predicts everything that Newton got right, but also predicts effects that happen close to massive objects, where gravity is very strong. Einstein didn't think of gravity as a force; he considered it the bending of space and time by the very presence of a massive object, such as a star. I get all bent out of shape just thinking about it.

Newton's concept of gravity explains the following:

>> Why the Moon orbits Earth, why Earth orbits the Sun, why the Sun orbits the center of the Milky Way, and why many other objects orbit one object or another out there in space

>> Why a star or a planet is round

>> Why gas and dust in space may clump together to form new stars

Einstein's theory of gravity, called the General Theory of Relativity, explains everything that Newton's theory does plus the following:

>> Why stars visible near the Sun during a total eclipse seem slightly out of position

>> Why black holes exist

>> Why gravitational lensing is found when we observe deep space

>> Why Earth drags warped space and time around with it as it turns, an effect that scientists have verified with the help of satellites orbiting Earth

>> How a collision of two black holes produces gravitational waves that shake things up even billions of light-years away

You find out about black holes in Chapters 11 and 13, and you can read up on gravitational lensing in Chapters 11, 14, and 15 without mastering the General Theory of Relativity.

You'll get smarter if you read every chapter in this book, but your friends won't call you Einstein unless you let your hair grow, parade around in a messy old sweater, and stick out your tongue when they take your picture.

Space: A Commotion of Motion

Everything in space is moving and turning. Objects can't sit still. Thanks to gravity, other celestial bodies are always pulling on a star, planet, galaxy, or spacecraft. Some of us are self-centered, but the universe has no center.

For example, Earth

>> Turns on its axis — what astronomers call *rotating* — and takes one day to turn all the way around.

>> Orbits around the Sun — what astronomers call *revolving* — with one complete orbit taking one year.

>> Travels with the Sun in a huge orbit around the center of the Milky Way. The trip takes about 250 million years to complete once, and the duration of the trip is called the *galactic year.*

>> Moves with the Milky Way in a trajectory around the center of the *Local Group of Galaxies,* a couple of dozen galaxies in our neck of the universe.

>> Moves through the universe with the Local Group as part of the *Hubble Flow,* the general expansion of space caused by the Big Bang.

The *Big Bang* is the event that gave rise to the universe and set space itself expanding at a furious rate. Detailed theories about the Big Bang explain many observed phenomena and have successfully predicted some that hadn't been observed before the theories were circulated. (For more about the Big Bang and other aspects of the universe, check out Part 4.)

Remember Ginger Rogers? She did everything Fred Astaire did when they danced in the movies, and she did it all backward. Like Ginger and Fred, the Moon follows all the motions of Earth (although not backward), except for Earth's rotation; the Moon rotates more slowly, about once a month. And it performs its tasks while also revolving around Earth (which it does about once a month).

And you, as a person on Earth, participate in the motions of rotation, revolution, galactic orbiting, Local Group cruising, and cosmic expansion. You do all that while you drive to work, whether you know it or not. Ask your boss for a little consideration the next time you run a few minutes late.

Chapter **2**

Join the Crowd: Skywatching Activities and Resources

A stronomy has universal appeal. The stars have fascinated people everywhere from prehistoric times into the modern age. Early observations of the sky led to all sorts of theories about the universe and attributions of power and purpose to the movements of stars, planets, and comets. As you look up at the sky, hundreds of thousands of people worldwide watch with you. When it comes to skywatching, you're not alone. Many people, organizations, publications, websites, smartphone apps, and other resources are at your disposal to help you get started, to keep you going, and to help you participate in the great work of explaining the universe.

In this chapter, I introduce you to these resources and give you suggestions on how you can get started. The rest is up to you. So join in!

TIP

When you know the resources, organizations, facilities, and equipment that can help you enjoy astronomy more deeply, you can move comfortably to the science of astronomy itself — the nature of the objects and phenomena out there in deep space. I describe the equipment you need to get started in Chapter 3.

You're Not Alone: Astronomy Clubs, Websites, Smartphone Apps, and More

You have plenty of readily available information, organizations, people, and facilities to help you get started and remain active in astronomy. You can join associations and activities to help researchers keep track of stars and planets. You can attend astronomy club meetings, lectures, and instructional sessions, which allow you to share telescopes and viewing sites to enjoy the sky with others. You also can find magazines, websites, books, computer programs, and smartphone apps with basic information on astronomy and current events in the sky.

Joining an astronomy club for star-studded company

The best way to break into astronomy without undue effort is to join an astronomy club. Clubs hold meetings where old hands pass along tips on techniques and equipment to beginners and scientists present talks. Members likely know where to get a good deal on a used telescope or binoculars and which products on the market are worth buying. (See Chapter 3 for more.)

Even better, astronomy clubs sponsor observing meetings, usually on weekend nights and occasionally on special dates during a meteor shower or another special event. An observing meeting is the best place to find out about the practice of astronomy and the equipment you need. You don't have to bring a telescope; most folks are happy to give you a look through theirs. Just wear sensible shoes, bring mittens and a hat for the cool night air, and put on a smile!

If you live in an urban area, chances are good that your night sky is bright. You can find better observational conditions if you travel to a dark spot in the country. Your local club probably has a good observing site, and when the members converge on that lonely place, you can enjoy safety in numbers.

TIP

If you live in a good-size city or a college town, you can probably find an astronomy club nearby. If you live in the United States, find the club(s) near you with the locator form of the NASA Night Sky Network, at `nightsky.jpl.nasa.gov`. Enter a city name, and a calendar pops up with the current month's astronomy activities in that area.

You can also check out the website of America's "club of clubs," the Astronomical League, at `www.astroleague.org`. Browse the list of more than 240 member societies, arranged by state.

GAZING AROUND THE WORLD: A SAMPLING OF ASTRONOMY CLUBS

The Astronomical Society of the Pacific (www.astrosociety.org), with headquarters in San Francisco, publishes the quarterly digital magazine *Mercury* for amateurs. It holds an annual meeting that moves around the western United States and sometimes goes as far east as Boston or Toronto. The society also offers numerous educational materials in astronomy to teachers.

Do you live in Canada? The Royal Astronomical Society of Canada has 29 *Centres*, which is a fancy name for astronomy clubs. Professionals from the nearest university often get involved with the Centre's activities. To find a Centre near you, consult the RASC website, www.rasc.ca.

In the United Kingdom, the venerable British Astronomical Association, founded in 1890, is still going strong. Its website is britastro.org. And the Society for Popular Astronomy, billing itself as "Britain's brightest astronomical Society," features sky event news and tweets about planets, bright meteors, and more on its colorful website, www.popastro.com.

Most other countries have astronomy clubs, too. Astronomy is truly a universal passion.

For a more global approach, visit the *Sky & Telescope* website, at www.skyandtelescope.com (just click the "Clubs and Organizations" tab in the Community menu to find clubs and organizations worldwide). The site lists over 20 astronomy clubs in the state of Missouri, for example, and 9 organizations, including planetariums and an observatory, in the nation of New Zealand.

Checking websites, magazines, software, and apps

Finding out about astronomy is easy. You can choose from a wide range of resources, including websites, apps for smartphones and tablet computers, magazines, and desktop computer software. The following sections offer some tips for finding the best information.

Traveling through cyberspace

The Net offers sites on every topic in astronomy, and the resources are increasing at, well, an astronomical rate! You can find many websites listed throughout this book; if you want more information on planets, comets, meteors, or eclipses, the web offers good sites on every topic.

TIP

The editors of *Sky & Telescope* magazine maintain one of the best websites, at www. skyandtelescope.com. Get your observational career started by checking out "This Week's Sky at a Glance" on that site. It gives a well-illustrated, day-by-day (or night-by-night) account of planets, comets, and other current space phenomena.

In the United Kingdom, *Astronomy Now* magazine offers a website with an "Observing" section that posts regular announcements of easily seen sky phenomena. Visit astronomynow.com.

Perusing publications

You can purchase excellent magazines to expand your knowledge of astronomy and your skill at practicing it. Most amateur astronomers subscribe to at least one publication. In many cases, if you join a local astronomy club, you may have access to a subscription to a national magazine at a member discount. (See "Joining an astronomy club for star-studded company" earlier in this chapter for the scoop on clubs.)

TIP

I recommend that you pick up a copy of each of the "big two" (literally, the biggest two) astronomical magazines: *Sky & Telescope* and *Astronomy*. Test-drive the publications for a month, and if you get more out of one than the other, go ahead and subscribe. You can do so from their websites at www.skyandtelescope.com and www.astronomy.com. Both of these magazines are available in both printed and digital editions.

Canadian readers can get the bimonthly *SkyNews: The Canadian Magazine of Astronomy & Stargazing*, a slick, full-color publication. Visit www.skynews.ca to subscribe.

Astronomy buffs in the United Kingdom should look for *Astronomy Now* and *Popular Astronomy* to see which magazine they prefer. Check their websites at astronomynow.com and www.popastro.com/popularastronomy.

In France, an excellent and well-illustrated magazine is *Ciel & Espace* (www. cieletespace.fr); in Australia, look for *Australian Sky & Telescope* and its astronomy yearbook for that country (www.austskyandtel.com.au). In Germany, *Sterne und Weltraum* (www.sterne-und-weltraum.de) excels.

Wherever you live, you'll find that the annual *Observer's Handbook* of the Royal Astronomical Society of Canada (www.rasc.ca) is very useful. Dozens of experts compile the handbook to help you enjoy the skies.

Surveying software and apps

A planetarium program, or "desktop planetarium," for your personal computer is a real plus. So is a planetarium app for your smartphone or tablet computer. Such programs or apps show you what the sky looks like from your home every night. You can also use them to find out what stars and planets will be up in the sky at a future date or at a different location so you can check in advance what you may observe on an upcoming vacation or visit to a dark sky observation site. This software is terrific to look at before you step outside to view the night sky. Some astronomers use these programs to plan their observing sessions. They prepare schedules of objects to scan with telescopes and binoculars at different times of the night to use their "dark time" effectively. Amateurs with certain telescope models that feature computer control can use some planetarium programs to guide their telescopes to stars, planets, or other sky objects of interest.

Desktop planetarium programs are available over a wide price range (including free programs) with many different features. You can find some programs advertised in astronomy and science magazines and on websites (see the previous two sections); they're updated occasionally for increased usefulness. You need only one program to get started, and that one may be the only program you ever need. The best way to select the planetarium program that suits you is to talk to experienced amateur astronomers at your local astronomy club. What works for them likely will work for you.

TIP

I recommend that you start out with Stellarium as the personal planetarium program on your desktop or laptop computer. It's a free, open-source program available for computers running most operating systems. It shows you the night or day sky at your place and time, or you can adjust it to check the sky at a later date. Visit the Stellarium website at www.stellarium.org to learn about its many features, view sample screen shots, or download the program to your computer.

A great many astronomy-related apps for smartphones and tablet computers are available for your consideration. Here are some that might work well for you:

>> **CraterSizeXL:** Use this iPad and iPhone app to calculate the possible danger if a potentially hazardous asteroid is headed for Earth (I discuss PHAs in Chapter 7). Fill in the available information on the object, and CraterSizeXL predicts the impact energy in units of Hiroshima-equivalent atom bomb blasts, crater size, and more. Damage from an asteroid hit can be in the trillions of dollars, but the good news is that you can download the app for about a buck.

>> **Sky Guide:** This award-winning app by Fifth Star Labs creates beautiful maps of the sky, complete with musical accompaniment (which you can mute). Walk outdoors and spot some stars that you don't recognize? Just turn on the app,

press Sky Guide's compass icon, and point the phone at the stars. Sky Guide (available for iPhone, iPad, and even for Apple Watch) displays a map of the region, names the constellation, draws lines between stars to indicate its shape, and prints the names of the brighter stars alongside them. It also shows you where the planets are and does much more. Check out the developer's website at www.fifthstarlabs.com.

>> **Galaxy Zoo:** This app is free for Android and Apple phones and tablets. It's for citizen scientists who help advance the science of astronomy by classifying the shapes of an astronomical number of galaxies that the Hubble Space Telescope (and others) have photographed. Join more than a quarter million volunteers worldwide in this worthy effort. (I describe galaxies and how to join Galaxy Zoo with your home computer, smartphone, or tablet in Chapter 12.)

>> **Google Sky Map:** If you have an Android phone or tablet, you can use this free app to identify visible stars and planets or to enjoy images of numerous celestial objects from NASA and other sources.

>> **GoSatWatch:** iPhone and iPad owners can use this app to learn where artificial satellites are orbiting and to predict when satellites will pass over your location (or any other) on Earth. (I describe artificial satellite observing in Chapter 4.) Satellite Safari is a similar app for both Android and Apple devices.

>> **SkySafari 5:** This highly rated planetarium program (for Android devices, iPhone, and iPad) is available in different versions at prices ranging from about $1 for the simplest version to about $20 for the most advanced. The more you pay, the more features you get. With the basic version, you can point the phone toward the sky, night or day, and it identifies the celestial objects visible (night) or invisible (day) that are up in that direction. Start with the cheap version and see whether it does everything you need.

>> **Star Chart:** This free iPhone and Android app is a simple way to identify stars and constellations.

Visiting Observatories and Planetariums

You can visit *professional observatories* (organizations that have large telescopes staffed by astronomers and other scientists for use in studying the universe) and *public planetariums* (specially equipped facilities with machines that project stars and other sky objects in a darkened room, accompanied by simple explanations of sky phenomena) to find out more about telescopes, astronomy, and research programs.

Ogling the observatories

You can find dozens of professional observatories in the United States and many more abroad. Some serve as research institutions operated by colleges and universities or government agencies. The U.S. Naval Observatory is in the heart of Washington, D.C., and has high security, so you must arrange tours in advance; they're usually on Monday nights (details are at www.usno.navy.mil/USNO). Some facilities are on remote mountaintops, such as the University of Denver's Mt. Evans Meyer-Womble Observatory, billed as the "Highest Operating Observatory in the West," at 14,148 feet; learn more at mysite.du.edu/~rstencel/MtEvans. (At the time of this book's publication, the observatory was not operational because it was still recovering from disastrous high winds that damaged the dome and the telescope during the 2011–2012 winter season.)

Certain observatories are dedicated to public education and information; cities, counties, school systems, or nonprofit organizations often operate these facilities. Here are some of the top sites:

>> **The Royal Observatory Greenwich, in London, England:** One of the most famous observatories in the world, at one time the Royal Observatory (www.rmg.co.uk/royal-observatory) was a professional research facility, then called the Royal Greenwich Observatory. The observatory is "home" of the prime meridian, from which longitude is measured around Earth. Equally important, it was the original source of Greenwich Mean Time, which set the standard for timekeeping worldwide.

>> **Lowell Observatory, on Mars Hill in Flagstaff, Arizona:** Research observatories vary in the degree to which they accommodate visits from the public, but Lowell is especially welcoming. You can even look at planets or stars on some nights: The observatory advertises that you can "Peer through the telescope that Percival Lowell used to sketch Mars or visit the telescope that helped Clyde Tombaugh discover Pluto." (However, the latter telescope was removed from its dome in 2017 for about a year's worth of renovations.) Lowell has a fine visitor center with a theater and exhibit hall and offers frequent tours. Check out its website at www.lowell.edu. (I discuss Percival Lowell and Mars in Chapter 6 and Pluto in Chapter 9.)

>> **The National Solar Observatory, at Sunspot, New Mexico:** This observatory runs Sun-watching telescopes in Lincoln National Forest above the little town of Cloudcroft, which is high above the city of Alamogordo. You can check out its Sunspot Astronomy and Visitor Center (daytime visits only) and take a tour of the observatory as well. See nsosp.nso.edu/pr for more.

>> **Mount Wilson Observatory, in the San Bernardino Mountains above Los Angeles, California:** Mount Wilson, where the expansion of the universe and the magnetism of the Sun were discovered, is a landmark in the history of

science. Albert Einstein was a special guest there, but you don't have to be a Big Brain to enjoy your visit. Self-guided tours are free; guided tours occur, for a fee, on Saturdays and Sundays. For a (high) price, you and your friends can even pool your money and book viewing time on the 100-inch telescope, which Edwin Hubble used to discover that the universe extends far beyond the Milky Way. See www.mtwilson.edu.

» **The Griffith Observatory, in Los Angeles, California:** This observatory is operated entirely for the public at Griffith Park in Los Angeles, offers planetarium shows and nighttime sky viewing, and is well worth visiting. Check out www.griffithobservatory.org.

» **Palomar Observatory, near San Diego, California:** Here you can see the famous 200-inch telescope that, for decades, was the largest and best in the world. Now, with new instrumentation, the telescope is still a great contributor of astronomical knowledge. You have to guide yourself around the observatory grounds. Check the website at www.astro.caltech.edu/palomar just before you visit; the facility closes well before sunset and is often off limits on short notice due to road and weather conditions near the summit. The observatory has a small museum and gift shop.

» **Kitt Peak National Observatory, on a Native American reservation (the Tohono O'odham Nation) in the Sonoran desert, 56 miles west of Tucson, Arizona:** When I worked at Kitt Peak during the 1960s, tourists were permitted to visit only during the day. Even so, there was a lot to see, from the visitor center (an astronomical museum) to the many telescope domes. Things have changed now, so put KPNO high on your list when you visit the American Southwest. During the day, you can take part in a guided or self-guided tour. You can also experience nighttime observing with certain KPNO telescopes. You must make an appointment for observing in advance. For details, go to www.noao.edu/kpno and click "Visiting Kitt Peak." Important: You will likely pass through a border control checkpoint on your way to Kitt Peak, and foreign visitors will need to show their passports.

» **MMT Observatory, on Mount Hopkins in the Coronado National Forest, 37 miles south of Tucson, Arizona:** Check in at the Fred Lawrence Whipple Visitor Center at the base of the mountain, where you can enjoy exhibits and sign up for tours of the observatory, with its 256-inch (6.5-meter) reflecting telescope, one of the largest in the continental United States. But before you go, check the MMT public tours website at www.mmto.org/node/289 to find when the visitor center is open, when tours are offered, and what they cost.

» **Mauna Kea Observatories, on the Big Island of Hawaii:** The biggest "telescope farm" in the United States is home to 13 large telescopes operated by the United States and other nations. (The only comparable assemblage of big telescopes on Earth is the European Southern Observatory in Chile.)

Mauna Kea is well worth visiting, but you need to be in good health, due to the high altitude (13,796 feet), and follow the "Visiting the Summit" instructions at the website (www.ifa.hawaii.edu/mko). The first time I visited the observatory, I walked up one flight of stairs and began to turn blue. My hosts administered oxygen and drove me down to a staff facility at lower altitude to recover. There's also a visitor information station at 9,200 feet, where you should start your visit and where you can participate in the nightly stargazing program.

You can also visit radio astronomy observatories, where scientists "listen" to radio signals from the stars or even seek signals from alien civilizations. Here are my top picks:

>> **The National Radio Astronomy Observatory's Very Large Array (VLA) in New Mexico:** Drive across the Plains of San Agustin, near Socorro, to see the huge radio telescope array, a system of 27 dish-shape radio telescopes, each 82 feet in diameter. Jodie Foster filmed scenes for the movie *Contact* here. Check out the visitor center and gift shop, but if you want a tour, you have to come on the first Saturday of the month. Tour times and other details are at www.vla.nrao.edu/. You pay a modest fee unless you're fortunate enough to attend one of the twice-yearly open houses.

>> **Green Bank Observatory:** At Green Bank, West Virginia, nestled among mountains in the United States National Radio Quiet Zone, you can tour the 328-foot Robert C. Byrd Green Bank Telescope, which is the world's largest fully steerable dish antenna. The Green Bank Science Center offers interactive exhibits, and it's the starting point for public tours (a fee applies). After taking in the sights, stop by the Starlight Café to treat yourself to tasty refreshments. Get more info at greenbankobservatory.org/visit/science-center.

>> **Jodrell Bank Observatory, near Goostrey, Cheshire, England:** At this observatory, operated by the University of Manchester, you can see the historic 250-foot Lovell Telescope, a big dish once used to bounce radar signals off Soviet booster rockets and even off the Moon. The Jodrell Bank Discovery Centre offers a Space Pavilion, Planet Walk, Galaxy Maze, and a large clockwork *orrery* (mechanical model of the solar system, with moving planets). It's closed on some holidays, so check the website (www.jodrellbank.net/visit/plan-a-visit/) before you go.

>> **Parkes Radio Telescope, near Parkes, New South Wales, Australia:** The radio telescope to visit when you're Down Under is the 210-foot, dish-shaped Parkes Radio Telescope. The telescope is known to astronomers for research findings, but it achieved its greatest public notice when it relayed radio transmissions to NASA from Apollo astronauts on their missions to the Moon. The Parkes Radio Telescope Visitors Discovery Centre has exhibits, a 3-D theater, and the aptly named Dish Café. See the website at www.atnf.csiro.au/outreach/visiting/parkes/index.html.

Popping in on planetariums

Planetariums, also called *planetaria*, are just right for beginning astronomers. They provide instructive exhibits and project wonderful sky shows indoors on the planetarium dome or on a huge screen. And many offer nighttime skywatching sessions with small telescopes, usually held outside in the parking lot, in an adjacent small observatory dome, or at a nearby public park. Many have excellent shops where you can browse the latest astronomy books, magazines, and star charts. The planetarium staff can also direct you to the nearest astronomy club, which may even meet after hours in the planetarium itself.

I practically grew up in the Hayden Planetarium at the American Museum of Natural History in New York City. Occasionally, I confess, I even snuck in for free. The planetarium staff was nice enough to have me back to speak (also for free) at its 50th anniversary. Although the old planetarium was torn down, a fine new one has replaced it. Make this planetarium, which is part of the Rose Center for Earth and Space, a prime destination when you visit the Big Apple. It's pricey but still much cheaper than a Broadway show, and its stars never miss a cue or sing off key (just don't try to sneak in like me)! Visitor information is on the American Museum's website at www.amnh.org/plan-your-visit.

TIP

You can find a list of planetariums in the United States, with links to their websites, at www.go-astronomy.com/planetariums.htm. To search for planetariums worldwide, consult the almost-400-page directory maintained by the International Planetarium Society. You can find it on the Society's website, www.ips-planetarium.org/?page=dir.

Vacationing with the Stars: Star Parties, Eclipse Trips, Dark Sky Parks, and More

An astronomy vacation is a treat for the mind and a feast for the eyes. Plus, traveling with the stars is often cheaper than taking a conventional holiday. You don't have to visit the hottest tourist destinations to keep up with your snooty neighbors. You can have the experience of a lifetime and come back raving about what you saw and did, not just what you ate and spent.

However, you can blow big bucks on one type of astronomy vacation: the eclipse cruise. But if you like ocean cruises, taking one to an eclipse doesn't cost any more than making a similar voyage that has no celestial rewards. Bargain-basement eclipse tours are available, too. Star parties, telescope motels, and visits to dark sky parks are additional options; I cover all the bases in the following sections. Pack your bags and have the neighbors watch the dog!

Party on! Attending star parties

Star parties are outdoor conventions for amateur astronomers. They set up their telescopes (some homemade and some not) in a field, and people take turns skywatching. (Be prepared to hear plenty of "Oohs" and "Ahs.") Judges choose the best homemade telescopes and equipment, earning their owners esteem and sometimes even a prize. If rain falls in the evening, partygoers may watch slide shows in a nearby hall or a big tent. Arrangements vary, but often some attendees camp in the field; others rent inexpensive cabins or commute from nearby motels. Star parties may last for a night or two, or sometimes as long as a week. They attract a few hundred to a few thousand (yes, thousand!) telescope makers and amateur astronomers. And the larger star parties have websites with photos of previous events and details on coming attractions. Some resemble the AstroFests I mention later in this section, with exhibitors and distinguished speakers as well as stargazing.

TIP

The leading star parties in the United States include

>> **Stellafane:** This Vermont star party has been going strong since 1926 (stellafane.org).

>> **Texas Star Party:** Commune with the stars on the mile-high Prude Ranch in the Lone Star State (texasstarparty.org).

>> **RTMC Astronomy Expo:** This worthy event takes place at Camp Oakes near Big Bear City at 7,600 feet in California's San Bernardino Mountains (www.rtmcastronomyexpo.org).

>> **Enchanted Skies Star Party:** Head to the desert for dark sky observing near Magdalena, New Mexico, and fine speakers (enchantedskies.org).

>> **Nebraska Star Party:** This party boasts "a fantastic light pollution–free sweep of the summer night sky" (www.nebraskastarparty.org).

Here are some of the leading star parties in the United Kingdom:

>> **The LAS Equinox Sky Camp:** Held at Kelling Heath, Norfolk, this party bills itself as "the largest star party in the U.K." (las-skycamp.org).

>> **Kielder Star Camp:** This twice-yearly event in the Northumberland International Dark Sky Park occurs in a forest thought to be "the darkest venue for any English star party" (sites.google.com/a/richarddarn.com/kielder-forest-star-camp-bookings/).

If you live in or plan to visit the Southern Hemisphere, check out these star parties:

>> **South Pacific Star Party:** It's held near Ilford, NSW, Australia, on a property reserved for skywatching by the Astronomical Society of New South Wales (www.asnsw.com/node/712).

>> **Central Star Party:** Try this party on New Zealand's North Island, a bit off the beaten path (www.censtar.party).

In the long run, I recommend that you visit at least one of these star parties, but in the meantime, you can ask at a local astronomy club meeting about a similar, although perhaps smaller, event that may be planned in your own area.

Getting festive at an AstroFest

You can hear famed astronomers, meet science authors, and learn the latest space news at an expo for astronomy enthusiasts. Astronomy groups organize the expos to reach out to the public, students, and educators. The events, often called Astro-Fests, feature talks on astronomy and space research, as well as displays and demonstrations of the latest technology available to amateur astronomers. The expos are most common under the AstroFest name in Europe and Australia. Search for one near you with your favorite search engine. Many of the large star parties (which I describe in the preceding section) in the United States also feature day-time speakers and exhibitors, as well as nighttime observing.

Tapping into Astronomy on Tap

Nearly all astronomy activities described in this book welcome children, but the Astronomy on Tap events are held in bars and aimed exclusively at adults. They consist of lectures and the like, accompanied by liquid refreshments and sometimes pub games such as Astronomy Trivia. Check the website at astronomyontap.org for an Astronomy on Tap near you, drink responsibly (if at all) when you attend, and consider bringing a designated driver who will not imbibe. As of 2016, there were 20 Astronomy on Tap programs worldwide, and the number was growing (or flowing).

To the path of totality: Taking eclipse cruises and tours

Eclipse cruises and tours are planned voyages to the places where you can view total eclipses of the Sun. Astronomers can calculate long in advance when and where an eclipse will be visible. The locations where you can see the total eclipse are limited to a narrow strip across land and sea, the *path of totality.* You can stay home and wait for a total eclipse to come to you, but you may not live long enough to see more than one, if any. So if you're the impatient stargazing type, you may want to travel to the path of totality.

Recognizing reasons to book a tour

If an eclipse is visible within easy driving distance, you don't need to sign up for a tour. (But such cases are rare; see the list of upcoming total eclipses of the Sun in Chapter 10, Table 10-1.)

If you're an experienced domestic and international traveler, you can go on your own to the path of totality for a distant eclipse. But consider this fact: Expert meteorologists and astronomers identify the best viewing locations years in advance. More often than not, these places aren't vast metropolises that offer huge numbers of vacant accommodations. You have to travel to random spots on the globe. After experts tab a spot as a prime location for a coming eclipse, tour promoters and savvy individuals book all or most of the local hotels and other facilities years in advance. Johnnies-or-Janies-come-lately, especially those who travel on their own, may be out of luck.

A tour promoter usually engages a meteorologist and a few professional astronomers (sometimes even me). So you have the benefit of a weatherperson to make last-minute decisions on moving the group's observing site to a place with a better next-day forecast, an astronomer to show you the safest methods to photograph the eclipse, and usually another lecturer who tells old eclipse tales and reports on the latest discoveries about the Sun and space.

On the night following the eclipse, viewers show their videos of the darkening sky, the birds coming to roost in the middle of the day, a klutz knocking over his telescope at the worst possible time, and the excited crowd saying "Wow" and "Hurray." And, of course, they replay the eclipse — over and over.

If these details haven't convinced you to take an eclipse tour for your viewing pleasure, consider this: Taking a group tour to a foreign destination is often cheaper than going on your own (and more satisfying than waiting years until a total eclipse is visible near your home). In the group, you'll make new friends who share your passion for eclipse chasing and skywatching. It's certainly happened to me.

Examining the advantages of cruising

An eclipse cruise is usually much better than a tour but more expensive. At sea, the captain and navigator have "2° of freedom." When the meteorologist says, "Head southwest down the path of totality for 200 miles" on the night before the eclipse (because of a late-breaking forecast for a cloud-free location at eclipse time), the ship can follow those instructions. But on land, you have to keep the bus on the road, and a road may not go in the direction you need. At a total eclipse in Libya, our procession of tour buses headed off the nearest road, proceeding across the desert to the designated viewing site, where water, port-a-pottys, security guards, and T-shirt vendors were on hand. On a cruise, you can leave the steering to the crew, recline in a deck chair, sip a beverage, have your camera ready, and wait for totality.

TIP

I've viewed many eclipses, and my experience is that if you stay on land, you get a clear view of totality about half the time. But if you travel on an ocean liner, you almost never miss.

Making the right decision

You can find advertisements for eclipse tours and cruises in astronomy and nature magazines, and on astronomy and travel websites Astronomy clubs, fraternal organizations, and college alumni societies often organize group accommodations on eclipse cruises.

TIP

Here are some ways you can choose the right tour or cruise for you:

>> **Consult current and back issues of astronomy magazines.** Most run articles about the viewing prospects for a solar eclipse a few years in advance. Get their expert recommendations about the best viewing sites.

>> **Check out the advertisements from travel operators.** Which tours and cruises go to the best places? Get brochures from travel agents, tour promoters, and cruise lines. Promoters often list previous successful eclipse trips, indicating a level of experience.

Motoring to telescope motels

Telescope motels are establishments where the attractions are the dark skies and the opportunity to set up your own telescope in an excellent viewing location. They usually have telescopes of their own that you can use, perhaps at an additional fee. If you fancy enjoying a stargazing vacation without lugging your own equipment across the country or the world, telescope motels are a good option.

YOU CAN DO IT! PARTICIPATING IN SCIENTIFIC RESEARCH

You can make your astronomy hobby beneficial and fun by joining national and world-wide efforts to gather precious scientific data. You may have only a pair of binoculars compared to Keck Observatory's two 10-meter-wide (400-inch) telescopes, but if the observatory has a cloudy day, Keck can't see anything. And if a spectacular fireball shoots over your hometown, you may be the only astronomer to see it.

Secret U.S. Department of Defense satellites and an amateur moviemaker vacationing at Glacier National Park recorded one of the most spectacular and interesting meteors of all time. A clip from that home movie appears in just about every scientific documentary about meteors, asteroids, and comets that appears on television. It pays to be in the right place at the right time. And someday you may be in that position.

Join other amateur astronomers in so-called citizen science and enjoy the projects I recommend throughout this book. You can help planetary geologists identify small surface features on Mars in images from spacecraft, hunt for hot stars in galaxies photographed by the Hubble Space Telescope, and even help map the Milky Way with data from two other NASA satellites. You may assist physicists who are searching for ripples in space-time called gravitational waves and aid the Search for Extraterrestrial Intelligence (SETI). The projects available to you change with time; most can be found at www.zooniverse.org. All you need is a computer with Internet access and some intelligence of your own.

TIP

Check out these telescope motels worth visiting in the United States:

>> **The Observer's Inn:** Located in the historic gold-mining town of Julian, California, this hotel has an observatory and concrete pads on which guests can place their own telescopes. Check it out online at www.observersinn.com.

>> **Primland:** Situated in the Blue Ridge Mountains near Meadows of Dan, Virginia, Primland no mere motel; it's a luxury resort that features stargazing in its own observatory (see primland.com). Invite your rich uncle and suggest that it's his treat.

>> **Furnace Creek Resort:** Venture into Death Valley National Park in California, where conditions can be ideal for stargazing. Bring your own telescope or attend a star party there conducted by the Las Vegas Astronomical Society. Find information at www.furnacecreekresort.com/activities/stargazing.

Overseas, telescope motels worth visiting include the following:

>> **AstroAdventures:** In the United Kingdom, consider a stay at this facility in North Devon. It offers an observatory, two lodges, Wi-Fi, and (for seasonal use) a swimming pool. See www.astroadventures.co.uk/ for more info.

>> **Carlo Magno Hotel Spa Resort:** Visit this facility in the Italian ski village of Madonna di Campaglio and observe with the hotel telescope, guided by a PhD astronomer from the University of Heidelberg. Look it up at www.hotelcarlomagno.com/en/hotel/astronomy.

>> **COAA (Centro de Observação Astronómica no Algarve):** Located in southern Portugal, this facility features telescopes, several guest suites, and a radar system to observe meteors day and night. Check it out online at www.coaa.co.uk.

>> **Hakos Guest Farm:** Bright stars and dark skies above a remote desert highlight this outpost in Namibia. The observatory is close to the guest rooms. Study the website at www.hakos-astrofarm.com/hakos_e.htm before you go.

>> **SPACE (San Pedro de Atacama Celestial Explorations):** This telescope motel in San Pedro de Atacama, Chile, offers lodging, tours (guides speak English, Spanish, and French), and telescopes on the high Atacama desert at an elevation near 8,000 feet. (See www.spaceobs.com.) The region is considered one of the finest locations for astronomical observatories on Earth.

Setting up camp at dark sky parks

The International Dark-Sky Association (IDA, darksky.org) confers the designation of International Dark Sky Park on public lands with good starry skies and limited or very low interference from artificial lighting. A dark sky park may or may not have telescopes of its own, but it's a place you can visit for a good view of the sky and to set up your own portable telescope.

Here are dark sky parks worth your visit in the United States:

>> **Natural Bridges National Monument:** This dark sky park in Utah claims that at night a bridge "forms a window into a sky filled with thousands of stars bright enough to cast a shadow." For more, see www.nps.gov/nabr/index.htm.

>> **Big Bend National Park:** This dark sky park is in Texas, on the Rio Grande. You can find it online at www.nps.gov/bibe/index.htm.

>> **Geauga County Observatory Park:** This Ohio park is furnished with telescopes and weather and seismic stations (www.geaugaparkdistrict.org/parks/observatorypark.shtml).

>> **Cherry Springs State Park:** The Susquehannock State Forest in Pennsylvania hosts this dark sky park, often the site of star parties (www.dcnr.state.pa.us/stateparks/findapark/cherrysprings).

>> **Clayton Lake State Park:** Watch the New Mexico night sky at Lake Observatory or attend sky talks by astronomers (www.emnrd.state.nm.us/SPD/claytonlakestatepark.html).

>> **Goldendale Observatory State Park:** This dark sky park in Washington State features an observatory, tours, and stargazing (parks.state.wa.us/512/Goldendale-Observatory).

>> **Headlands International Dark Sky Park:** View the night sky along the Straits of Mackinac, in northern Michigan (www.midarkskypark.org/).

>> **Stephen C. Foster State Park:** This Georgia attraction gives you entrée to the Okefenokee Swamp with its famous black water, but it has black skies too, recognized by designation as an International Dark Sky Park. The website (www.gastateparks.org/StephenCFoster) suggests that you "Join a ranger-guided paddle in a canoe or kayak while the sun sets over the Okefenokee." There's a fee and limited space in each boat.

In Europe, consider exploring these dark sky parks:

>> **Galloway Forest Park:** This dark sky location is the largest forest park in Scotland (www.gallowayforestpark.com).

>> **Sark Dark Sky Community:** In the Channel Islands off the coast of Normandy, Sark has no public street lights and no motor vehicles but farm tractors (darksky.org/idsp/communities/sark/).

>> **Exmoor National Park:** Located in the southwest of England, this park was the first designated International Dark Sky Reserve in Europe (www.exmoor-nationalpark.gov.uk/enjoying/stargazing).

>> **Hortobágy National Park:** This dark sky park in Hungary is situated in a landscape little altered since the last Ice Age (darksky.org/idsp/parks/hortobagy/).

>> **Lauwersmeer National Park:** Built on land reclaimed from the sea in the Netherlands, Lauwersmeer is a relatively dark area in a very light-polluted country (darksky.org/idsp/parks/lauwersmeer/).

In the Southern Hemisphere, you can visit Aoraki Mackenzie Dark Sky Reserve. This huge reserve on New Zealand's South Island extends over more than 1,600 square miles (darksky.org/idsp/reserves/aorakimackenzie/).

When you enjoy an international dark sky park, you'll recognize that "lights out" is a good thing.

Chapter **3**

Terrific Tools for Observing the Skies

f you've ever gone outside and looked at the night sky, you've been stargazing — observing the stars and other objects in the sky. Naked-eye observation can distinguish colors and the relationships between objects — like finding the North Star by using "pointer stars" in the Big Dipper.

From naked-eye observation, you can take a short step up by adding optics to see fainter stars and view objects with greater detail. First try binoculars, and then graduate to a telescope. Next thing you know, you're an astronomer!

But I'm getting ahead of myself. First, you need to take some quiet looks at the cosmos and see the beauty and mystery for yourself. You can use three basic tools, at least one of which you already own.

Whether you use your eyes, a pair of binoculars, or a telescope, each method of observation is best for some purposes:

>> **The human eye:** Your peepers are ideal for watching meteors, the aurora borealis, a planetary conjunction (when two or more planets are close to each other in the sky), or a conjunction of a planet and the Moon.

>> **Binoculars:** A good pair of binoculars is best for observing bright variable stars, which are too far from their *comparison stars* (stars of known constant brightness used as a reference to estimate how much another star varies in brightness) to see them together through a telescope. And binoculars are wonderful for sweeping through the Milky Way and viewing the bright nebulae and star clusters that dot it here and there. Some of the brighter galaxies — such as M31 in Andromeda, the Magellanic Clouds, and M33 in Triangulum — look best through binoculars.

>> **Telescope:** You need a telescope to get a decent look at most galaxies and to distinguish the members of close double stars, among many other uses. (A *double star* consists of two stars that appear very close together; they may or may not be near each other in space, but when they truly are together, they form a binary star system.)

In this chapter, I cover these observational tools, provide you with a quick primer on the geography of the night sky, and give you a handy plan for delving into astronomy. Before long, you'll be observing the skies with ease.

Seeing Stars: A Sky Geography Primer

When viewed from the Northern Hemisphere, the whole sky seems to revolve around the North Celestial Pole (NCP). Close to the NCP is the North Star (also called Polaris), a good reference point for stargazers because it always appears in almost exactly the same place in the sky, all night long (and all day long, but you can't see it then).

In the following sections, I show you how to familiarize yourself with the North Star and give you a few facts about constellations.

As Earth turns . . .

Our Earth turns. The Greek philosopher Heraclides Ponticus proclaimed that concept in the fourth century B.C. But people doubted Heraclides' observations because folks thought that, if his theories were true, they should feel dizzy like riders of a fast merry-go-round or a whirling chariot. They couldn't imagine a turning Earth if they didn't physically feel the effects. Instead, the ancients thought the Sun raced around Earth, making a complete revolution every day. (They didn't feel the effects of Earth's rotation, nor do you and I, because those effects are too small to notice.)

Proof of Earth's turning, or *rotation*, didn't come until 1851, more than two millennia after Heraclides (researchers didn't have much government funding back then, so progress was slow). The proof came from a big French swinger: a heavy metal ball suspended from the ceiling of the Meridian Room at the Paris Observatory, and then from the ceiling of a church in Paris (the Panthéon) on a 220-foot wire. The device is called a *Foucault pendulum*, after the French physicist who came up with the plan and carried out the first demonstration. If you kept an eye on the pendulum as it swung back and forth all day, you could see that the direction taken by the swinging ball across the floor gradually changed, as though the floor was turning underneath it. And it was — the floor turned with Earth.

TIP

If you're not convinced that Earth turns, or if you just like big swingers, you can see a 240-pound brass Foucault pendulum in the rotunda of the Griffith Observatory in Los Angeles (www.griffithobservatory.org/exhibits/centralrotunda_foucaultpendulum.html). The Oregon Convention Center in Portland claims the world's largest Foucault pendulum, named Principia and considered a work of art. You see why in a photo at www.oregoncc.org/visitors/public-art-collection. It's open to visitors; you don't have to be registered for a convention. In England, check out the University of Manchester's Foucault pendulum (www.mace.manchester.ac.uk/project/teaching/civil/structuralconcepts/Dynamics/pendulum/pendulum_pra3.php).

If you already believe that Earth turns, however, you can verify that conclusion as you just sip a favorite beverage and watch the Sun set in the west.

As I explain in Chapter 1, the rotation of Earth around its axis makes the stars and other sky objects appear to move across the sky from east to west. In addition, the Sun moves across the sky during the year on a circle called the *ecliptic*. (If you could see the stars in the daytime, you'd note the Sun moving to the west across the constellations, day by day.) The ecliptic is inclined by 23.5° to the celestial equator, the same angle by which the axis of Earth is tilted from the perpendicular to its orbital plane.

The planets stay close to the ecliptic as they move throughout the year. They move systematically through 12 constellations located on the ecliptic, which, collectively, are called the *Zodiac*: Aries, Taurus, Gemini, Cancer, Leo, Virgo, Libra, Scorpius, Sagittarius, Capricorn, Aquarius, and Pisces. The names of those constellations are what astrology buffs call the signs of the Zodiac. (Actually, a 13th constellation, Ophiuchus, intersects the ecliptic, but in ancient times, it wasn't included in the Zodiac, so it's not a sign.)

Earth's steady progression along its orbit of the Sun results in a different appearance of the night sky over the course of the year. (The same effect is what makes the Sun move around the circle of the ecliptic; as Earth moves, we see the Sun in different directions against the background of fixed stars.) The stars aren't located in the same places with respect to the horizon throughout the night or throughout a year (the North Star is an exception; it is in nearly the same place all night and every night). The constellations that appeared high in the sky at dusk a month ago are lower in the west after sunset now. And if you view the constellations that loom low in the east just before dawn, you preview what you'll see overhead at midnight in a few months.

TIP

To keep track of the constellations, use the sky maps that come monthly in astronomy magazines such as *Sky & Telescope* and *Astronomy*. (For more about magazines, check out Chapter 2.) And you can get an inexpensive *planisphere*, or star wheel, which features a rotating disk in a square frame, with a hole cut into it that represents the limits of your view. You want a planisphere designed for your latitude, or roughly so.

Here are three good planispheres to consider:

>> *The Night Sky* by David Chandler is manufactured in different sizes and for various latitudes. It's available in English-, Japanese-, and Spanish-language editions. Get the version that corresponds to your latitude, and pick the larger size if you have a choice. See www.davidchandler.com.

>> *David H. Levy Guide to the Stars* is colorful and designed for children. It's suitable for use in the United States and in other places that are between latitudes 30° and 60° north. At the time of this writing, it's a bargain at about $4 on Amazon.

>> *Star Wheel* is available from the Sky & Telescope Shop (www.shopatsky.com). This planisphere is published in four versions for various northern and southern latitudes.

Using a planisphere helps you understand the motions of the stars as you turn the disk to show the stars for different hours and dates. But if you don't care about learning that stuff and just want to know what constellations are up in the sky, you're better off downloading a suitable app for your smartphone or tablet computer, as I describe in Chapter 2. An app is especially helpful if you're on a trip to the opposite hemisphere because any of the constellations that you know from viewing them back home will look upside down, and other constellations of course will be new to you.

. . . keep an eye on the North Star

Anyone can walk outside on a clear night and see some stars. But how do you know what you're seeing? How can you find it again? What do you watch for?

One of the most time-honored ways of getting familiar with the night sky if you live in the Northern Hemisphere is to pay attention to the North Star, or Polaris, which barely moves. After you identify which way is north, you can orient yourself to the rest of the northern sky — or to anywhere else, for that matter. In the southern sky, you need to find the bright stars Alpha and Beta Centauri (a Southern Hemisphere planisphere, smartphone app, or simple star map can help), which point the way to the Southern Cross.

You can easily find the North Star by using the Big Dipper in the constellation Ursa Major (see Figure 3-1). The Big Dipper is one of the most easily recognized sky patterns. If you live in the continental United States, Canada, or the United Kingdom, you can see it every night of the year.

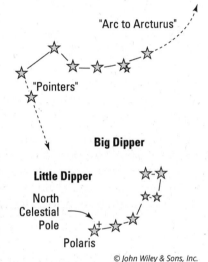

FIGURE 3-1: The Big Dipper points to other sights in the sky.

REMEMBER

The two brightest stars in the Big Dipper, Dubhe and Merak, form one end of its bowl and point directly to the North Star. The Big Dipper helps you locate the bright star Arcturus, in Bootes, too: Just imagine a smooth continuation of the curve of the Dipper's handle, as in Figure 3-1.

The stars close to Polaris never set below the horizon in the United Kingdom, nor at most latitudes in North America, which makes them *circumpolar stars:* They appear to circle around Polaris. Ursa Major is a circumpolar constellation seen from almost all the Northern Hemisphere. The circumpolar area of the sky depends on your latitude. The closer you live to the North Pole, the more of the sky is circumpolar. And in the Southern Hemisphere, the farther south your location, the greater the circumpolar part of the sky. But if a constellation is circumpolar in the

Northern Hemisphere, it cannot be circumpolar in the Southern Hemisphere, and vice versa.

Orion is a distinctive constellation visible in the Northern Hemisphere's winter evening sky, with the three stars that make up its belt pointing in one direction to Sirius in Canis Major and in the opposite direction to Aldebaran in Taurus. Orion also contains the 1st magnitude stars Betelgeuse and Rigel, two brilliant beacons in the sky (see Figure 3-2). For more on the magnitude of stars, head to Chapter 1.

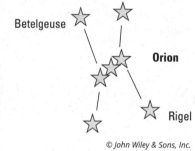

FIGURE 3-2:
Orion and its bright, beaconlike stars Rigel and Betelgeuse.

© John Wiley & Sons, Inc.

TIP

You can become friends with the night sky by looking at this book's constellation maps (check out Appendix A) and checking them out with your own eyes. Just as becoming familiar with the streets of your city helps you find your way faster, knowing the constellations helps you set your sights on the sky objects you want to observe. Gaining sky knowledge also assists you in tracking the appearance of the stars and their movements as you maneuver through a nightly session.

Beginning with Naked-Eye Observation

If you don't already know the compass directions in your area, take the time to familiarize yourself with them. You need to know north from south and east from west. When you get your bearings, you can use the weekly sky highlights from the *Sky & Telescope* website (www.skyandtelescope.com) or use a planisphere, a planetarium-type smartphone app, or a desktop planetarium program to orient yourself to the brightest stars and planets in the sky at the time you look. (For more about these astronomy resources, see Chapter 2 and the earlier section "Seeing Stars: A Sky Geography Primer.") When you recognize the bright stars, you have an easier time picking out the patterns of slightly fainter ones all around them.

Table 3-1 lists some of the brightest stars you can see in the night sky, the constellations that contain them, and their magnitudes (the measure of their brightness — see Chapter 1). Many of them are visible from the continental United States, the United Kingdom, and Canada. Some you can see well from only southern latitudes, so bright stars that you can't see in the United States may be prominent sights for Australians. See Chapter 11 for information on spectral class, which is an indication of the color and temperature of a star. (Spectral class B stars, for example, are white and rather hot, and M stars are red and relatively cool.)

TIP

Start your observations by consulting one of the sky map resources I describe earlier in the chapter and see how many of the brightest stars you can locate at night. After you do, try identifying some of the dimmer stars in the same constellations. And, of course, keep your eye out for the bright planets: Mercury, Venus, Mars, Jupiter, and Saturn (which I cover in Chapters 6 and 8).

TABLE 3-1 **The Brightest Stars As Seen from Earth**

Common Name	Apparent Magnitude	Constellation Designation	Spectral Class
Sirius	–1.5	α Canis Majoris	A
Canopus	–0.7	α Carinae	A
Arcturus	–0.04	α Bootes	K
Rigil Kentaurus	–0.01	α Centauri	G
Vega	0.0	α Lyrae	A
Capella	0.1	α Aurigae	G
Rigel	0.1	β Orionis	B
Procyon	0.4	α Canis Minoris	F
Achernar	0.5	α Eridani	B
Betelgeuse	0.5	α Orionis	M
Hadar	0.6	β Centauri	B
Altair	0.8	α Aquilae	A
Aldebaran	0.9	α Tauri	K
Antares	1.0	α Scorpii	M
Spica	1.0	α Virginis	B
Pollux	1.1	β Geminorum	K
Fomalhaut	1.2	α Piscis Austrini	A
Deneb	1.3	α Cygni	A
Acrux	1.3	α Crucis	B

In winter and summer, the Milky Way runs high in the sky at most locations in the Northern Hemisphere. If you can recognize the Milky Way as a wide, faintly luminous band across the sky, you have at least a pretty fair observing site. And if the Milky Way seems bright, you must be at a dark sky location like those that I describe in Chapter 2.

TIP

The most important step in observing with the naked eye is to shield your vision from interfering lights. If you can't get to a dark place in the country, find a dark spot in your backyard or possibly on the roof of a building. You won't eliminate the light pollution high up in the sky that results from the collective lights of your city, but trees or the wall of a house can prevent nearby lights, including streetlamps, from shining in your eyes. After 10 or 20 minutes, you can see fainter stars; you're getting *dark adapted.*

Watching the beautiful comet Hyakutake in 1996 from a small city in the Finger Lakes of northern New York, I found that walking around the corner of a building to shield myself from nearby street lights made the comet a much better sight.

Ideally, you want a site with a good horizon and only trees and low buildings in the distance, but finding that kind of location is next to impossible in a major urban area.

REMEMBER

If you fail to find a site with a good horizon in all directions, the most important horizon is the southern one (if you live in the Northern Hemisphere). You make most observations in the Northern Hemisphere while facing roughly south (with east to the left and west to the right). As you face south, the stars rise to your left and set to your right. If you observe from the Southern Hemisphere, reverse this procedure and face north: The stars rise in the east (on your right) and set in the west (on your left).

TIP

Always have a watch, a notebook, and a dim or red flashlight to use for recording what you see. Some flashlights come with a red bulb, or you can buy red cellophane from a greeting card store to wrap around the lamp. After you become dark adapted, white light reduces your ability to see the faint stars, but dim red light doesn't hurt your dark adaptation.

When I was young, some amateur observers would dictate their observations into a portable tape recorder in real time. That way, they could keep their eyes on the sky — for example, while watching for meteors — and avoid the difficulty of writing things down in dim red light or with a cold wind blowing. Nowadays, you may be able to record your observations with a voice recorder app on your smartphone.

HOW BRIGHT IS BRIGHT?

I talk about magnitude in Chapter 1, but it helps to know that astronomers can define magnitude in different ways for different purposes:

- **Absolute magnitude** is the brightness of a sky object as seen from a standard distance of 32.6 light-years. Astronomers consider it the "true" magnitude of the object.

- **Apparent magnitude** is how bright an object appears from Earth, which is usually different than its absolute magnitude depending on how far away from Earth the sky object is located. A star closer to Earth may appear brighter than one farther away, even if its absolute magnitude is fainter.

- **Limiting magnitude** is the apparent magnitude of the faintest star that you can see. It depends on how clear the sky is at the time of observation and how dark the sky is. A very bright star may be invisible if many clouds hang overhead, for example, and city lights or a full Moon may interfere with viewing fainter stars that you can see with the naked eye under good conditions. Limiting magnitude is especially important in meteor and deep sky observations. On a clear, dark night, the limiting magnitude may be 6 at the zenith, but in the city, the limiting magnitude may be only 3 or 4.

Star charts depict the apparent magnitudes of the stars to simulate their appearance in the sky.

Using Binoculars or a Telescope for a Better View

As with any new hobby, you want to gain some experience and research what's available before you start buying expensive equipment. You don't want to buy a telescope until you've seen several telescopes of different types in action and discussed them with other observers. In the following sections, I offer my advice for choosing the right binoculars or telescope for you.

WARNING

Don't even think about looking at the Sun with a telescope or binoculars unless you use the safe procedures and special equipment that I mention in "Staying safe when you view the Sun" later in this chapter and describe in detail in Chapter 10. Otherwise, you can suffer permanent harm.

Binoculars: Sweeping the night sky

Owning a good pair of binoculars is a must. Buy or borrow a pair and observe with them before you get a telescope. Binoculars are excellent for many kinds of observation, and if you give up on astronomy (sigh), you can still use them for many other purposes. And if you borrowed the binoculars, be sure to return them before you are viewed with suspicion.

REMEMBER

Binoculars are great for observing variable stars, searching for bright comets and novae, and sweeping the sky just to enjoy the view. You may never discover a comet yourself, but you'll certainly want to view some of the brighter ones as they appear. Nothing works better for this purpose than a good pair of binoculars.

The following sections cover the way binoculars are specified according to their capabilities, and I show you the steps to take as you figure out what kind of binoculars to buy. Figure 3-3 takes you inside a pair of binoculars.

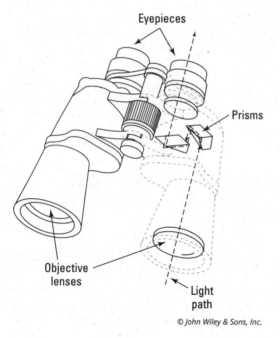

FIGURE 3-3:
Binoculars are like a pair of telescopes coordinated for your eyes.

Eyepieces

Prisms

Objective lenses

Light path

© John Wiley & Sons, Inc.

Prisms, glass, and shapes

Binoculars contain prisms to bend the light coming from the two large lenses (*objective lenses*) to the two smaller lenses (*eyepieces*) that you look through. It's necessary because the eyepieces can't be farther apart than the distances between your eyes, or you won't be able to look through both eyepieces at once. The objectives are bigger than your eyes, so they need to be farther apart; therefore, the light paths through the binoculars have to be bent.

The two basic prism types in binoculars are as follows:

>> **Roof prisms** are used in binoculars that are relatively straight and narrow; these binoculars are favorites among bird-watchers.

>> **Porro prisms** are used in binoculars that are relatively wide and short; they are the better type for stargazing because they give brighter images for the same-size lenses. It's also easier to hold wide binoculars steady.

Binoculars use two main types of glass:

>> **BK-7 glass,** a trade term for garden-variety borosilicate glass, is often used in cheap binoculars.

>> **BaK-4 glass,** or barium crown glass, is used in fine binoculars and often yields brighter images of dim astronomical objects.

Deciphering the numbers on binoculars

Binoculars come in many sizes and types, but each pair of binoculars is described by a numerical rating — 7×35, 7×50, 16×50, 11×80, and so on. (Note that the ratings read "7 by 35, 7 by 50, 16 by 50, and 11 by 80." Don't say "7 times 35.") Here's how to decode these ratings:

>> The first number is the optical magnification. A 7×35 or 7×50 pair of binoculars makes objects look seven times larger than they do to the naked eye.

>> The second number is the *aperture,* or diameter, of the light-collecting lenses (the big lenses) in the binoculars, measured in millimeters. An inch is about 25.4 millimeters; thus, 7×35 and 7×50 binoculars have the same magnifying power, but the 7×50 pair has bigger lenses that collect more light and show you fainter stars than the 7×35 pair.

Also keep the following considerations in mind:

>> Bigger binoculars reveal fainter objects than smaller ones do, but they weigh more and are harder to hold and point steadily toward the sky.

>> Higher-magnification binoculars, such as 10×50 and 16×50, show objects with greater clarity, provided that you can hold them steady enough, but they have smaller fields of view, so finding celestial targets is harder than with lower-magnification binoculars.

>> Giant binoculars — 11×80, 20×80, and on up — are heavy and hard to hold steady; many people can't use them without a tripod or stand. You can use some really big binoculars only with a heavy stand that may come with them, and they cost thousands of dollars. They're definitely not for beginners.

>> Many intermediate sizes are available, such as 8×40 or 9×56.

TIP

Here's my opinion: 7×50 is the best size for most astronomical purposes and certainly the best size to start with. If you purchase binoculars much smaller than 7×50, you really equip yourself for bird-watching rather than astronomy. Most astronomers can use 7×50 binoculars without a tripod or stand, although some people may need to brace themselves to hold these binoculars steady. Buy a much larger size than 7×50, and you may be investing in a white elephant that you'll rarely use.

Making sure your binoculars are right for you

First and foremost, don't buy binoculars unless you can return them after a trial run. Here's how to make the basic check to determine whether a pair of binoculars is worth keeping:

>> The image should be sharp across the field of view when you look at a field of stars.

>> You should have no difficulty focusing the binoculars for your eyesight, with a separate adjustment for at least one of the eyepieces (the small lenses next to your eyes when you look through the binoculars).

>> When you adjust the focus, it should change smoothly. Stars' images should be sharp points when in focus and circular in shape when not.

>> Special transparent coatings are deposited on the objective lenses (large lenses) of many binoculars. This feature, called *multicoating,* results in a clearer, more contrasting view of star fields. Binoculars that are *fully multicoated* are even better; in them, the coatings are applied to all the lenses and prisms.

TIP

Some astronomers wear their eyeglasses when observing with binoculars. Others, like me, are more comfortable putting away the glasses when using binoculars. But if you don't have your eyeglasses on, you may have a problem writing notes, reading a star chart, and so on. Your choice of the best binoculars for you may depend on whether you will keep your glasses on, as I explain next.

If you plan to wear your eyeglasses when you observe with binoculars, you need binoculars that have enough *eye relief.* Eye relief is the distance (measured in millimeters) from the outer surface of a binocular eyepiece to the *focal point,* where the binoculars focus an image. If your eye is beyond that distance from the eyepiece, you can't see the whole field of view. That circumstance happens when the thickness of your eyeglasses keeps your eye from getting to the focal point. Here's my advice: Disregard salesperson assurances and binoculars manufacturers' specs on eye relief. Instead, perform this simple test on binoculars that you are considering buying:

1. **Take off your glasses and focus the binoculars on a scene a block away (or in the sky).**

   Note how much of the scene is in view.

2. **Put on your glasses.**

   If the eyepieces have rubber eye cups, fold down the cups so that you can get closer to the eyepieces while wearing glasses.

3. **Focus the binoculars on the same scene as before.**

   If you don't see as much of the scene with your glasses on, the binoculars don't have enough eye relief.

WARNING

Good binoculars are sold in optical and scientific specialty stores. Some large camera stores have decent binocular selections. But I recommend avoiding department stores. You may get low-grade merchandise in some department stores or pay exorbitant prices for fancy binoculars in others. And you can bet that the salespeople peddling them know less than you do.

You can pay hundreds of dollars or even a few thousand bucks for a good pair of 7×50 binoculars, but if you shop around, you can find a perfectly adequate pair for $120 or less. (Pawn shops and military surplus stores are excellent places to look.) Used binoculars are often a good deal, but you must try them before you buy them because they may be out of whack.

Many astronomers buy their binoculars from specialty retailers and manufacturers that advertise in astronomy magazines and on the web (see Chapter 2 for more about these resources). If you must order your binoculars online or by mail, first ask experienced amateurs you meet at an astronomy club or consult a staff member at a planetarium to find a reliable dealer.

Reputable makers of binoculars include Bushnell, Canon, Celestron, Fujinon, Meade, Nikon, Orion, Pentax, and Vixen. Some high-end Canon and Nikon binoculars have image stabilization, a high-tech feature that makes the image much steadier. They come in handy on a boat rocking at sea and are often a great help on land as well.

Telescopes: When closeness counts

If you want to look at the craters on the Moon, the rings of Saturn, or Jupiter's Great Red Spot (all of which I describe in Part 2), you need a telescope. The same advice goes for observing faint variable stars or viewing all but the brightest galaxies or the beautiful small glowing clouds called *planetary nebulae*, which have nothing to do with planets (see Chapters 11 and 12).

WARNING

Before you view the Sun or any object crossing in front of the Sun, however, be sure to read the special instructions in Chapter 10 to protect your eyes and avoid going blind!

The following sections cover telescope classifications, mounts, and shopping tips to find the best telescope for your needs.

Focusing on telescope classifications

Telescopes come in three main classifications:

>> *Refractors* use lenses to collect and focus light (see Figure 3-4). In most cases, you look straight through a refractor.

>> *Reflectors* use mirrors to collect and focus light (see Figure 3-5). Reflectors come in different types:

- In a *Newtonian* reflector, you look through an eyepiece at right angles to the telescope tube.

- In a *Cassegrain* telescope, you look through an eyepiece at the bottom.

- A *Dobsonian* reflector gives you the most *aperture* (or light-gathering power) for your money, but you may have to stand on a stool or a ladder to look through it. Dobsonians tend to be larger than other amateur telescopes (because large Dobsonians are more affordable), and the eyepiece is up near the top.

>> *Schmidt-Cassegrains* and *Maksutov-Cassegrains* use both mirrors and lenses. These models are more expensive than reflectors with comparable apertures, but they're compact and more easily taken on observing trips.

Many varieties are available within these general telescope types. And every telescope used for amateur purposes is equipped with an *eyepiece*, which is a special lens (actually, a combination of lenses mounted together as a unit) that magnifies the focused image for viewing. When you take photographs, you usually remove the eyepiece and mount the camera on the telescope in its place.

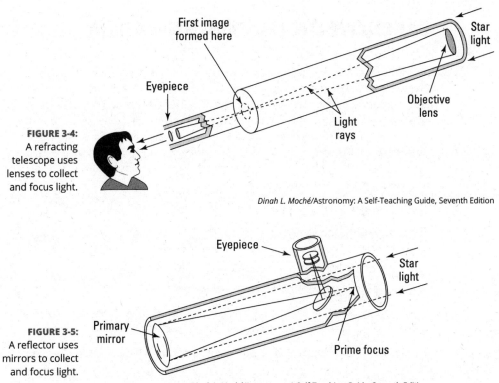

FIGURE 3-4:
A refracting telescope uses lenses to collect and focus light.

FIGURE 3-5:
A reflector uses mirrors to collect and focus light.

Just as with a microscope or a camera with interchangeable lenses, you can use interchangeable eyepieces with almost any telescope. Some companies don't make telescopes at all, specializing instead in making eyepieces that work with many different telescopes.

TIP

Beginners usually buy the highest-magnification eyepieces they can, which is a great way to waste your money. I recommend low- and medium-power eyepieces because the higher the power, the smaller the field of view, making it tougher to track faint (and possibly even bright) targets. For a small telescope, observation is usually best with eyepieces that give magnifications of 25x or 50x, not 200x or more. (The x stands for "times," as in 25 times bigger than what you see with the naked eye.) If you see a telescope advertised for its "high power," the advertisers may be trying to sell mediocre goods to unwitting buyers. And if a salesperson touts the high power of a telescope, patronize another store.

WARNING

What limits your view of fine details with a small telescope isn't the power of the eyepiece. Air turbulence (the same factor that makes stars seem to twinkle) and any shaking of the telescope in the breeze are the effects that limit the clarity of your view.

CALCULATING THE MAGNIFICATION OF YOUR EYEPIECE

Eyepieces are usually marked with their focal lengths in millimeters, and a given eyepiece may provide different magnifications with different telescopes. If you know the focal length of your telescope, you can calculate the magnification that a particular eyepiece provides. Follow these steps to determine the magnification your eyepiece provides for your telescope:

1. **Determine the focal length of the telescope.**

   If your telescope lists the focal length, make note of it and skip to Step 3.

   To calculate the focal length, multiply the f/number listed on the telescope, such as f/5.6 or f/8, by the diameter of the main lens or mirror. For example, if the diameter is 6 inches and the f/number is f/8, the focal length is $6 \times 8 = 48$ inches.

2. **Convert the focal length from inches to millimeters.**

   Multiply the length in inches by 25.4 (the number of millimeters per inch). For example, if your focal length is 48 inches, $48 \times 25.4 = 1,219.2$ millimeters.

3. **Divide the focal length of your telescope by the focal length of the eyepiece.**

   If the telescope focal length is 1,219.2 millimeters and the eyepiece focal length is 25 millimeters, the eyepiece provides a magnification of approximately 49 times $(1,219.2 \times 25 = 48.768)$.

Examining telescope mounts

Telescopes are generally mounted on a stand, a tripod, or a pier in one of two ways:

>> With an *alt-azimuth mount,* you can swivel the telescope up and down and side to side — in *altitude* (the vertical plane) and *azimuth* (the horizontal plane). You need to adjust the telescope on both axes to compensate for the motion of the sky as Earth rotates. Dobsonian reflectors always use alt-azimuth mounts.

>> With the more expensive *equatorial mount,* you align one axis of the telescope to point directly at the Celestial North Pole or, for Southern Hemisphere viewers, the Celestial South Pole. After you spot an object, simply turning the telescope around the polar axis keeps it in view. Be sure to polar-align the scope for each viewing session.

COLORING YOUR UNIVERSE

What do you see when you look at a sky object through your binoculars or telescope? Do you see glorious stars, planets, and sky objects in brilliant colors, as shown in the photos in the color section of this book? Not!

Sorry for the tease, but you're likely to see most stars and sky objects in pale colors. Most stars look white or off-white by eye, binoculars, or telescope — yellowish rather than yellow, for example. The colors are most vivid when adjacent stars have contrasting colors, as found in telescopic views of some double stars.

Most photos of sky objects have been color enhanced, traditionally described as having *false color*. Astronomers don't use false color to gussy up the universe, which is fine as it is. Nor is it meant to give a false impression of the deep sky. In fact, the enhancement furthers the search for truth, much like a stain on medical slides brings out the detail in the cells and helps identify physical differences and relationships.

Depending on the method of observation and presentation, photos of the same object can be strikingly different. But they all tell scientists about differences in the structure of the object, what substances it may contain, and what dynamic processes are taking place. Also, many astronomical images are obtained in forms of light invisible to the human eye (like radio waves, ultraviolet, infrared, and X-ray), so astronomers use false color in this absence of any recognizable color.

An alt-azimuth mount is usually steadier and easier for a beginner to use, but an equatorial mount is better for tracking the stars as they rise and set. However, if a telescope has built-in computer control (as in both models that I recommend shortly), either style of mount is fine because the computer takes care of tracking.

The objects you see in the telescope are usually upside down, which isn't the case with binoculars. Of course, it doesn't really make much difference in the viewing, but just know that top and bottom are reversed when you view through a telescope. Adding a lens to rotate the image so it's right side up has the disadvantage that it reduces the amount of light coming through the scope and dims the image, so it's best not done. When viewed through an equatorially mounted telescope, a star field maintains the same orientation as it rises and sets. But when seen through an alt-azimuth mount telescope, the field rotates during the night, so the stars on top wind up on the side.

STAYING SAFE WHEN YOU VIEW THE SUN

Taking even the briefest peek at the Sun through a telescope, binoculars, or any other optical instrument is dangerous unless the device is equipped with a solar filter made by a reputable manufacturer specifically for viewing the Sun. And the filter must be properly mounted on the telescope, not jury-rigged.

You must also use a solar filter when you view Mercury as it crosses *(transits)* the disk of the Sun. (I describe transits of Mercury in Chapter 6.) Viewing an object against the Sun requires the use of protective viewing techniques because you're also looking at the Sun. If you have a Newtonian reflector, a Dobsonian reflector, or a refractor, you can try using projection. See Chapter 10 for specifics about solar viewing and how to protect your eyes.

Shopping for telescopes the smart (and economical) way

WARNING

A cheap, mass-manufactured telescope, often called a *drugstore* or *department store* telescope, is usually a waste of money. And it still costs more than a hundred or maybe several hundred dollars.

A good telescope, bought new, may run you several hundred dollars to $1,000, and you certainly can pay more. But you can find alternatives:

>> Used telescopes are often sold through ads in astronomy magazines or in the newsletters of local astronomy clubs. If you can inspect and test a used telescope and you find what you like, buy it! A well-maintained telescope can last for decades.

>> In many areas, amateurs can observe with the larger telescopes operated by astronomy clubs, planetariums, or public observatories.

The technology of amateur telescopes is advancing at a rapid pace, and a former astronomer's dream can be today's obsolete equipment. Quality and capabilities are going up, and prices are generally fair, perhaps because reliable manufacturers are in competition for your money.

Generally, a good refractor gives better views than a good reflector that has the same *aperture*, or telescope size. Aperture refers to the diameter of the main lens, the mirror, or, in a more complicated telescope, the size of the unobstructed portion of the optics. But a good refractor is more expensive than a comparable reflector.

The Maksutov-Cassegrains and Schmidt-Cassegrains are good compromises between the low cost of a reflector and the high performance but high cost of a refractor. For many astronomers, these hyphenated telescope types are the preferred models.

One of the best small telescopes is the Meade ETX-90. Its aperture is 3.5 inches, almost the smallest size of any telescope that you should start with. (If you find a good instrument at a good price from a 2.5-inch aperture on up, especially in a refractor, consider it for purchase.)

The ETX-90 sells for about $400 and comes with an Autostar computerized controller and a tripod. This instrument automatically points at almost any object you specify if that object is in view from your location at that time. The Autostar can even find moving objects, such as planets, based on stored information, and it's equipped to give you a "tour" of the best sights in the sky, selected with no input from you.

A good competing telescope for the ETX-90 is the Celestron SkyProdigy 90. It's comparably sized and equipped and has the capability to automatically align itself on the sky, after which it points to almost any celestial object that you select. It goes for about $600.

You definitely don't want to spend this much money on a telescope until you see the same model in action at an astronomy club observing meeting or a star party (see Chapter 2). But the price is no more than you pay for a fine camera and an accessory lens or two. You can find larger telescopes for less money — check the ads in current issues of astronomy magazines — but you have to invest much more effort in learning to use them effectively.

Some brand-name telescopes are sold through authorized dealers that tend to have expert knowledge. But take their advice with just a wee bit of salt.

Here are some key websites to browse for telescope product information:

>> Celestron, for many years the favorite manufacturer for thousands of astronomers (www.celestron.com)

>> Meade Instruments Corporation (www.meade.com)

>> Orion Telescopes & Binoculars (www.telescope.com)

On each of these websites, you can find the instruction manuals for many of the telescopes that they sell. Consider taking a look at the manual before you buy a telescope so you know whether it will be helpful when you run into problems.

GOOD SEEING GONE BAD

Turbulence in the atmosphere affects how well you can see the stars. It makes the stars seem to twinkle. The term *seeing* describes the conditions of the atmosphere relating to steadiness of the image — *good seeing* is when the air is stable and the image holds steady. You often have better seeing late at night, when the heat of the day has dissipated. When seeing is bad, the image tends to "break up," and double stars blur together in telescopic views. The stars always twinkle most close to the horizon, where the seeing is worst.

The bright planets Mercury, Venus, Mars, Jupiter, and Saturn generally don't twinkle when you view them with the naked eye. They are not single points of light like a star seen from Earth, but are seen as disks. Each disk consists of many individual points of light. The separate points twinkle but average each other out (one twinkling brighter, another twinkling fainter at any moment), so you see a steady light from the planet.

The warmth of a telescope brought out from a heated home into the cool night air causes some bad seeing; wait a while for the telescope to cool down, and your viewing will improve. Situations vary, but 30 minutes is usually enough to make a significant difference in the viewing.

If you don't live in the United States, you may be able to find dealers for the telescope brands in your country. The Widescreen Centre in London carries Celestron, Meade, and Orion telescopes (www.widescreen-centre.co.uk). In Australia, the Binocular and Telescope Shop, with locations in Sydney and Melbourne, is among the dealers that carry these telescopes (www.bintel.com.au).

Planning Your First Steps into Astronomy

I recommend that you get into the astronomy hobby gradually, investing as little money as possible until you're sure about what you want to do. Here's a plan for acquiring both basic skills and the needed equipment:

1. **If you have a late-model computer, invest in a free or inexpensive planetarium program.**

    Better yet, if you have a smartphone, download and use a free or cheap planetarium app (see recommendations for apps and programs in Chapter 2). Start making naked-eye observations at dusk on clear nights and before dawn, if you're an early riser.

To plan your observations of planets and constellations, you can also rely on the weekly sky scenes at the *Sky & Telescope* website (www.skyandtelescope.com). If you don't have a suitable computer, plan your observations based on the monthly sky highlights in *Astronomy* or *Sky & Telescope* magazine.

2. **After a month or two of familiarizing yourself with the sky and discovering how much you enjoy it, invest in a serviceable pair of 7×50 binoculars.**

3. **As you continue to observe the bright stars and constellations, invest in a star atlas that shows many of the dimmer stars, as well as star clusters and nebulae.**

    Sky & Telescope's Pocket Sky Atlas by Roger W. Sinnott (Sky Publishing, 2007) is a good choice. For maps that are equally good but larger, consult the *Jumbo Pocket Sky Atlas* by the same author and publisher (2016); you'll just need a bigger pocket. Compare scenes in your star atlas with the constellations that you're observing; the atlas shows their RAs and Decs (see Chapter 1 for info about RAs and Decs.) Eventually, you'll start to develop a good feel for the coordinate system.

4. **Join an astronomy club in your area, if at all possible, and get to know the folks who have experience with telescopes (see Chapter 2 for more about finding clubs).**

5. **If all goes well and you want to continue in astronomy — as I bet you will — invest in a well-made, high-quality telescope in the 2.5-to-4-inch size range.**

    Study the telescope manufacturer websites earlier in this chapter or send for catalogs advertised in astronomy magazines. Better yet, talk to experienced astronomy club members if you can. They can advise you on buying a new telescope, and they may know someone who wants to sell a used telescope.

TIP

You may be able to borrow a starter telescope and try it out at home. Thanks to the New Hampshire Astronomical Society (NHAS), a movement to place such telescopes in public libraries has begun. Astronomy clubs purchase the telescopes; club members modify them for use by inexperienced borrowers and then donate them to the libraries. The telescope model adopted for this project is the Orion StarBlast 4.5, which retails for about $210. It's meant for use on a tabletop but may work for you when just placed on the ground. According to *Sky & Telescope*, by late 2016 NHAS had placed over 100 of these telescopes in New Hampshire libraries, and the St. Louis Astronomical Society had placed over 130 in Missouri and Illinois libraries. Astronomy clubs in other areas are beginning to sponsor library telescopes; search the web to see whether a library telescope program exists near you. Who knows; you may have a (star) blast!

If you find that you enjoy astronomy as much as I think you will, after a few years, consider moving up to a 6- or 8-inch telescope. It may be harder to use, but you'll be ready to master it after you have some experience. Equipped with a larger telescope, you can see many more stars and other objects. You can get ideas about what larger telescopes to consider by talking to other amateur astronomers and by attending a star party, where you can see many different telescopes in operation and on display. (I cover star parties in Chapter 2.)

Chapter **4**

Just Passing Through: Meteors, Comets, and Artificial Satellites

See a moving object in the daytime sky? You probably know whether it's a bird, a plane, or Superman. But in the night sky, can you distinguish a meteor from a flash of light glinting off an Iridium satellite? And among objects that move slowly but perceptibly across the starry background, can you tell a comet from an asteroid?

This chapter defines and explains many of the objects that sweep across the night sky. (The Sun, Moon, and planets move across the sky, too, but in a more stately procession. I focus on them in Parts 2 and 3.) When you know how to identify these night visitors, you can look forward to enjoying them all.

Meteors: Wishing on a Shooting Star

No astronomy term is misused more often than the word *meteor*. Amateur astronomers and even scientists are quick to spurt out *meteor* when *meteoroid* or *meteorite* is the accurate term. Take a look at the correct meanings:

» A *meteor* is the flash of light produced when a naturally occurring small, solid object (a meteoroid) enters Earth's atmosphere from space; people often call meteors "shooting stars" or "falling stars."

» A *meteoroid* is a small, solid object in space, usually a fragment from an asteroid or comet, that orbits the Sun. Some rare meteoroids are actually rocks blasted off Mars and Earth's Moon.

» A *meteorite* is a solid object from space that has fallen to the surface of Earth. About 100 tons of meteoritic material fall on Earth every day (some estimates are even higher).

If a meteoroid runs into Earth's atmosphere, it may produce a meteor bright enough for you to see. If the meteoroid is big enough to hit the ground instead of disintegrating in midair, it becomes a meteorite. Many people hunt for and collect meteorites because of their value to scientists and collectors.

The two main kinds of meteoroids have different places of origin:

» *Cometary meteoroids* are fluffy little dust particles shed by comets.

» *Asteroidal meteoroids,* which range in size from microscopic particles to boulders, are literally chips from asteroids — the so-called minor planets — which are rocky bodies that orbit the Sun (and which I describe in Chapter 7).

When you go to a science museum and see a meteorite on display, you're examining an asteroidal meteoroid that fell to Earth (or, in rare cases, a rock that fell after being knocked off the Moon or Mars by a larger impacting body). It may be made of stone, iron (actually, an almost rustproof mixture of nickel and iron), or both. Showing rare simplicity (for once), scientists call these meteorite types, respectively, *stony, iron,* and *stony-iron meteorites.*

In the following sections, I cover three types of meteors: sporadic meteors, fireballs, and bolides. I also give you the scoop on meteor showers.

TIP

For reliable information on how to observe, record, and report meteors, visit the "Meteor" section of the British Astronomical Association website at `www.britastro.org/section_front/19`. While you're surfing the web, head over to the International Meteor Organization's website at `www.imo.net` for the latest news and photos of fireballs (which I explain in the following section) and galleries of meteor pictures and videos that IMO members have uploaded to the site.

COMB YOURSELF FOR SPACE DUST

If an astronomer finds a *micrometeorite* (a meteorite so small that you must view it through a microscope), it may be a particle that began as a cometary meteoroid, or it may be a very small asteroidal meteoroid.

Micrometeorites are so small that they don't create enough friction to burn up or disintegrate in the atmosphere, so they sift slowly down to the ground. Chances are, you have one or two pieces of this space dust in your hair right now, but the dust is almost impossible to identify because it would be lost among the millions of other microscopic particles on your head (no offense).

Scientists obtain micrometeorites by flying ultraclean collecting plates on high-altitude jet aircraft. And they drag magnetized rakes, which pick up micrometeorites made of iron, through the mud on the sea floor. If you collect enough gunk from gutters on the roofs of buildings, you can find micrometeorites there too. You just have to accumulate hundreds of pounds of the debris and search through it with advanced laboratory techniques. Investigators who sifted through 660 pounds of matter from rooftops in Oslo, Norway, and Paris, France, struck pay dirt: They ended up with 48 certifiable micrometeorites for their pains. So if anyone ever accuses you of stooping to the gutter, tell them you're proud to be a meteorite hunter.

On January 2, 2004, NASA's Stardust space probe flew past Comet Wild-2 (a comet that passes inside the orbit of Mars once every six years or so and is fairly easy to reach with a probe) and collected some comet dust. The probe ended its "wild ride" by ejecting a capsule containing the dust it had collected. It parachuted down in Utah on January 15, 2006. A team of 200 scientists analyzed the tiny particles. They found that some dust grains came from other stars, but most formed near the Sun, some so close that, according to Stardust scientist Donald Brownlee, it was "hot enough to evaporate bricks." The probe, renamed Stardust-NeXT, flew on to photograph Comet Tempel 1 on Valentine's Day in 2011. You can see many of the photos at `stardustnext.jpl.nasa.gov/`.

Spotting sporadic meteors, fireballs, and bolides

When you're outdoors on a dark night and see a "shooting star" (the flash of light from a random, falling meteoroid), what you're probably seeing is a *sporadic* meteor. But if many meteors appear, all seeming to come from the same place among the stars, you're witnessing a *meteor shower.* Meteor showers are among the most enjoyable sights in the heavens; I devote the next section in this chapter to them.

A dazzlingly bright meteor is a *fireball.* Although a fireball has no official definition, many astronomers consider a meteor that looks brighter than Venus to be a fireball. However, Venus may not be visible at the time you see the bright meteor. So how can you decide whether you're seeing a fireball?

TIP

Here's my rule for identifying fireballs: If people facing the meteor all say "Ooh" and "Ah" (everyone tends to shout when they see a bright meteor), the meteor may be just a bright one. But if people who are *facing the wrong way* see a momentary bright glow in the sky or on the ground around them, it's the real thing. To paraphrase an old Dean Martin tune, when the meteor hits your eye like a big pizza pie, that's a fireball!

Fireballs aren't very rare. If you watch the sky regularly on dark nights for a few hours at a time, you'll probably see a fireball about twice a year. But *daylight fireballs* are very rare. If the Sun is up and you see a fireball, mark it down as a lucky sighting. You've seen one tremendously bright fireball. When nonscientists see daytime fireballs, they almost always mistake them for an airplane or missile on fire and about to crash.

TIP

Any very bright fireball (approaching the brightness of the half Moon or brighter) or any daylight fireball represents a possibility that the meteoroid producing the light will make it to the ground. Freshly fallen meteorites are often of considerable scientific value, and they may be worth good money, too. If you see a fireball that fits this description, write down all the following information so your account can help scientists find the meteorite and determine where it came from:

1. **Note the time, according to your watch.**

    At the earliest opportunity, check how fast or slow your watch is running against an accurate time source, such as the Master Clock at the U.S. Naval Observatory, which you can consult at www.usno.navy.mil/USNO. If you have a smartphone, it should give you the time accurate at least to the minute.

2. **Record exactly where you are.**

    If you have a Global Positioning System receiver handy (or a smartphone with a GPS app, such as Compass on the iPhone), take a reading of your latitude and longitude. Otherwise, make a simple sketch showing where you stood when you saw the fireball — note roads, buildings, big trees, or any other landmarks.

3. **Make a sketch of the sky, showing the track of the fireball with respect to the horizon as you saw it.**

   Even if you're not sure whether you faced southeast or north-northwest, a sketch of your location and the fireball track helps scientists determine the trajectory of the fireball and where the meteoroid may have landed.

After a daylight fireball or a very bright nighttime fireball, interested scientists advertise for eyewitnesses. They collect the information, and by comparing the accounts of persons who viewed the fireball from different locations, they can close in on the area where it most likely fell to the ground. Even a brilliant fireball may be only the size of a small stone — one that would fit easily in the palm of your hand — so scientists need to narrow the search area to have a reasonable chance of finding it. If you don't see a call for information after your fireball observation, chances are good that the nearest planetarium or natural history museum will accept your report and know where to send it. Or report your fireball observation to the American Meteor Society at www.amsmeteors.org — just look for the prominent "Report a Fireball" link on their home page.

A *bolide* is a fireball that explodes or produces a loud noise even if it doesn't break apart. At least, that's how I define it. Some people use *bolide* interchangeably with *fireball.* (You won't find an official agreement on this term; you can find different definitions in even the most authoritative sources.) The noise you hear is the sonic boom from the meteoroid, which is falling through the air faster than the speed of sound.

When a fireball breaks apart, you see two or more bright meteors at once, very close to each other and heading the same way. The meteoroid that produces the fireball has fragmented, probably from aerodynamic forces, just as an airplane falling out of control from high altitude sometimes breaks apart even though it hasn't exploded.

Often a bright meteor leaves behind a luminous track. The meteor lasts a few seconds or less, but the shining track — or *meteor train* — may persist for many seconds or even minutes. If it lasts long enough, it becomes distorted by the high-altitude winds, just as the wind gradually deforms the skywriting from an airplane above a beach or stadium.

REMEMBER

You see more meteors after midnight local time than before because, from midnight to noon, you're on the forward side of Earth, where our planet's plunge through space sweeps up meteoroids. From noon to midnight, you're on the backside, and meteoroids have to catch up in order to enter the atmosphere and become visible. The meteors are like bugs that splatter on your auto windshield. You get many more on the front windshield as you drive down the highway than on the rear windshield because the front windshield is driving into bugs and the rear windshield is driving away from bugs.

Watching a radiant sight: Meteor showers

Normally, only a few meteors per hour are visible — more after midnight than before and (for observers in the Northern Hemisphere) more in the fall than in the spring. But on certain occasions every year, you may see 10, 20, or even 50 or more meteors per hour in a dark, moonless sky far from city lights. Such an event is a *meteor shower*, when Earth passes through a great ring of billions of meteoroids that runs all the way around the orbit of the comet that shed them. (I discuss comets in detail later in this chapter.) Figure 4-1 illustrates the occurrence of a meteor shower.

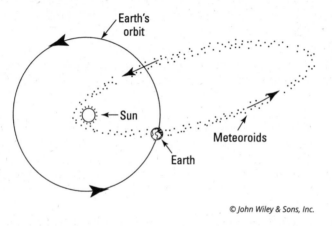

FIGURE 4-1:
Earth's path crossing a belt of meteoroids creates a shower of meteors.

© John Wiley & Sons, Inc.

The direction in space or place on the sky where a meteor shower seems to come from is called the *radiant*. The most popular meteor shower is the Perseids, which, at its peak, produces as many as 80 meteors per hour. (The Perseids get their name because they seem to streak across the sky from the direction of the constellation Perseus, the Hero, their radiant. Meteor showers are usually named for constellations or bright stars [such as Eta Aquarii] near their radiants.)

A few other meteor showers produce as many meteors as the Perseids, but fewer people take the time to observe them. The Perseids come in August, when the balmy nights in North America and Europe often are perfect for skywatching, but the other leading meteor showers — the Geminids and Quadrantids — streak across the sky in December and January, respectively, when the weather is worse in the Northern Hemisphere and observers' ambitions are limited.

Table 4-1 lists the top annual meteor showers. The dates in the table are the nights when the showers usually reach their peak. Some showers go on for days, and others for weeks, raining down meteors at lower rates than the peak values. The Quadrantids may last for just one night or only a few hours.

TABLE 4-1   **Top Annual Meteor Showers**

Shower Name	Approximate Date	Meteor Rate (Per Hour)
Quadrantids	Jan. 3–4	90
Lyrids	Apr. 21	15
Eta Aquarids	May 4–5	30
Delta Aquarids	July 28–29	25
Perseids	Aug. 12	80
Orionids	Oct. 21	20
Geminids	Dec. 13	100

The Quadrantids' radiant is in the northeast corner of the constellation Bootes, the Herdsman. The meteors are named for a constellation found on 19th-century star charts that astronomers no longer officially recognize. In addition to losing their namesake, the Quadrantids seem to have lost the comet that spawned them — their origin was a mystery until 2003, when astronomer Petrus Jenniskens found that an object named 2003 EH_1 may be their parent comet.

The Geminids are a meteor shower that seems to be associated with the orbit of an asteroid rather than a comet. However, the "asteroid" is probably a dead comet, which no longer puffs out gas and dust to form a head and tail. The object 2003 EH_1, the likely parent of the Quadrantids, may be a dead comet, too. (I discuss comets in the next section.)

The Leonids are an unusual meteor shower that occurs around November 17 every year, usually to no great effect. But every 33 years, many more meteors are present than usual, perhaps for several successive Novembers. Huge numbers of Leonids were seen in November 1966 and again in November 1999, 2000, 2001, and 2002, at least for brief times at some locations. The next great display likely will come in 2032. Don't forget to look for it.

You almost never see as many meteors per hour as I list in Table 4-1. The official meteor rates are defined for exceptional viewing conditions, which few people experience nowadays. But meteor showers vary from year to year, just like rainfall. Sometimes people do see as many Perseids as listed. On rare occasions, they see many more than expected. Such inconsistency is why keeping accurate records of the meteors that you count can be helpful to the scientific record.

TIP

For more information on upcoming meteor showers, check out the American Meteor Society website at www.amsmeteors.org.

Do you live south of the equator? If so, check out the list of meteor showers visible from the Southern Hemisphere on the website of the Royal Astronomical Society of New Zealand, at rasnz.org.nz/in-the-sky/meteor-showers.

To track meteors, you need an accurate time source, a way to record your observations, and a dim flashlight to see what you're doing.

TIP

The best light for astronomical observations is a red flashlight, which you can purchase, or make from an ordinary flashlight by wrapping red transparent plastic around the bulb. Some astronomers paint the lamp with a thin coat of red nail polish. If you use a white light, you dazzle your eyes and make it impossible to see the fainter stars and meteors for 10 to 30 minutes, depending on the circumstances. Letting your vision adjust to the dark is called getting *dark adapted* and is a step you want to take every time you observe the night sky.

TIP

The best way to watch and count meteors is to recline on a lounge chair. (You can do pretty well just lying on a blanket with a pillow, but you're more likely to fall asleep in that position and miss the best part of the show.) Tilt your head so you're looking slightly more than halfway up from the horizon to the zenith (see Figure 4-2) — the optimum direction for counting meteors. Take notes. And be sure you have a thermos of hot coffee, tea, or cocoa!

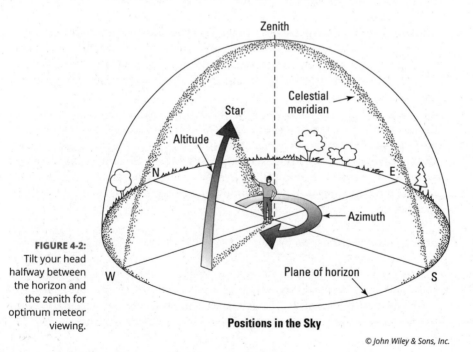

FIGURE 4-2:
Tilt your head halfway between the horizon and the zenith for optimum meteor viewing.

Positions in the Sky

© *John Wiley & Sons, Inc.*

You don't have to face the radiant when you observe a meteor shower, although many people do. The meteors streak all over the sky, and their visible paths may begin and end far from the radiant. But you can visually extrapolate the meteors' paths back in the direction from which they seem to come, and the paths point back to the radiant. Identifying a radiant in that way is how you can tell a shower meteor from a sporadic one.

If you do face the radiant, however, you see some meteors that seem to have very short paths, even though they appear fairly bright. The paths appear short because the meteors are coming almost right at you. Fortunately, the shower meteoroids are microscopic and won't make it to the ground.

NOT JUST A FLASH IN THE PAN

An extreme case of a daylight fireball and bolide occurred on February 15, 2013, when a large meteoroid flew over Chelyabinsk, Siberia, just after sunrise, breaking apart in midair and producing a blast that burst windows in the city. No one was hurt by the meteorite itself, but over 1,000 people, including schoolchildren, were injured, mostly by broken glass. Observers said the fireball was much brighter than the sun. When it hit Earth, the meteoroid became a meteorite, although few reporters made this technical distinction. Later, observers noticed a hole in the ice covering nearby Lake Chebarkul (not a small thing in Siberia in winter); eight months later, divers recovered the largest known remaining chunk of the meteorite from the bottom of the lake. It weighed about 1,440 pounds (653 kilograms), but the original object was much larger before it shattered.

The Chelyabinsk event was the most serious cosmic impact on Earth since June 30, 1908, when a much larger midair explosion toppled trees across an area of about 800 square miles in a remote Siberian forest. Fortunately, there were few people in the area and no known deaths. The disaster region, near the Stony Tunguska River, was so isolated that the first scientific investigator didn't arrive for eight years! Astronomers' opinions differ as to whether the impacting object in the so-called Tunguska event was a very large meteoroid (equivalent to a small asteroid) or a small comet. Either way, it shook the ground like an earthquake, detectable for thousands of miles, and apparently generated high-altitude atmospheric particles that caused a bright glow in the evening sky that was soon seen as far away as England. Whether meteoroid or comet, the Tunguska object came without warning and dealt a blow that could easily have destroyed a large urban area or more and killed hundreds of thousands of people had it struck a populous, developed part of Earth.

For the hard facts about asteroids, including the near-Earth objects that sometimes pose a threat to Earth, flip to Chapter 5.

PHOTOGRAPHING METEORS AND METEOR SHOWERS

Digital cameras are now the preferred tools for photographing meteors. But digital meteor photography requires a digital single lens reflex camera (DSLR), which is expensive camera (point-and-shoot cameras and cellphone cameras don't work very well, except in the rare case when you can catch a brilliant fireball) and a lot of trial-and-error experimenting until you get it right. Further, you need a DSLR that you can set for time exposures and that accepts a cable for an intervalometer or "remote switch with digital timer."

You might need to spend more on a suitable camera for meteor photography than on a decent small telescope for other observations, but the camera can be used for other purposes, not just your astronomy hobby.

Here are some important guidelines for digital meteor photography:

- Observe from as dark a location as possible, away from urban lighting.

- Try meteor photography only when the Moon is below the horizon.

- Use a sturdy tripod so the camera doesn't shake during a time exposure.

- Use a wide-angle lens (because you'll catch more meteors in a single shot than with a normal lens) and set it on Infinity. Don't use a telephoto lens.

- Use an intervalometer or "remote switch with digital timer" to operate the camera shutter without shaking the camera and to take pictures at regular intervals during the night.

- Point the camera about halfway up the sky from the horizon to the zenith, or a little higher, facing whichever direction has the least interfering sky glow from city or other lights.

- Spend some time making test exposures to determine what settings to use on that particular night. (The best settings vary depending on how bright the sky is.) Make several 10-second exposures, some 20-second exposures, and some 30-second exposures. You're trying to determine how long you can let an exposure last (the longer the better) without skylight overexposing the picture. You may need to repeat this series of time exposures for each of two or three ISO settings. (With a larger ISO setting, you can record fainter meteors, which means more meteors, but with the larger ISO setting, the sky overexposes sooner, so you can't expose for as long a time.) With experience, you should find the "sweet spot" of exposure time and ISO that works best with your lens at your location.

- For more info on digital meteor photography read the expert advice at www.amsmeteors.org/meteor-showers/how-to-photograph-meteors-with-a-dslr/.

You can photograph sporadic meteors by following the preceding guidelines, but there aren't many sporadic meteors to catch on any given night. A meteor shower offers you the opportunity to snap more meteors, as long as the Moon isn't in the sky. With moonlight, you'll catch far fewer meteors, if any. When photographing a meteor shower, take the photographs when the shower radiant (the constellation from which the meteor shower seems to come) is well above the horizon, preferably 40° or more. The horizon is at 0° altitude, and the zenith (overhead point) is 90° up, so the halfway point between them is at 45°; two-thirds of the way up is 60°, and so on.

TIP

For more information on meteor showers, including historical events, facts, and advice on observing, head to the Sky and Telescope site (www.skyandtelescope.com) and enter "Shooting Stars" in the search window. Then you can download the free *Shooting Stars* e-book (you may have to register your email address).

Comets: The Lowdown on Dirty Ice Balls

Comets, great blobs of ice and dust that slowly track across the sky looking like fuzzy balls trailing gassy veils, are popular visitors from the depths of the solar system. They never fail to attract interest. Every 75 to 77 years, the best-known ice ball, Halley's Comet, returns to our neck of the woods. If you missed its appearance in 1986, try again in 2061! If you're impatient, you can see other interesting comets in the meantime. Often a less famous comet, such as Hale-Bopp in 1997, is much brighter than Halley's.

TIP

Many people confuse meteors and comets, but you can easily distinguish them by these criteria:

>> A meteor lasts for seconds; a comet is visible for days, weeks, or even months.

>> Meteors flash across the sky as they fall overhead, within 100 miles or so of the observer. Comets crawl across the sky at distances of many millions of miles, often appearing almost motionless over the course of an hour or several hours unless you look through a telescope.

>> Meteors are common; comets that you can easily see with the naked eye come less than once a year, on average.

Astronomers believe that comets were born in the vicinity of the outer planets, starting near the orbit of Jupiter and extending well beyond Neptune. The comets near Jupiter and Saturn were gradually disturbed by the gravity of those mighty planets and flung far out into space, where they fill a huge, spherical region well beyond Pluto — the Oort Cloud — extending roughly 10,000 AU from the Sun. (One AU, or astronomical unit, is a distance equal to about 93 million miles.) Other comets were ejected to or were formed and remain in the Kuiper Belt (see Chapter 9), a region that starts around the orbit of Neptune and continues to a distance of about 50 AU from the Sun, or about 10 AU beyond Pluto. Passing stars occasionally disturb these regions and send comets on new orbits, which may take them close to Earth and the Sun, where we can see them.

In the following sections, I discuss a comet's structure, famous comets throughout time, and methods you can use to spot a comet.

Making heads and tails of a comet's structure

A comet is a stuck-together mixture of ice, frozen gases (such as the ices of carbon monoxide and carbon dioxide), and solid particles — the dust or "dirt" shown in Figure 4-3. Historically, astronomers described comets as having a head and tail or tails, but with additional research, they've been able to clarify the nature of a comet's structure.

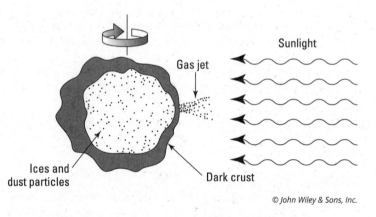

FIGURE 4-3:
A comet is really just a dirty ice ball.

© John Wiley & Sons, Inc.

The nucleus

Astronomers initially named a bright point of light in the head of a comet the *nucleus*. Today we know that the nucleus is the true comet — the so-called dirty ice ball. The other features of a comet are just emanations that stem from the nucleus.

A comet far from the Sun is only the nucleus; it has no head or tail. The ice ball may be dozens of miles in diameter or just a mile or two. That size is pretty small by astronomical standards, and because the nucleus shines only by the reflected light of the Sun, a distant comet is faint and hard to find.

Images of Halley's nucleus from a European Space Agency probe that passed very close to it in 1986 show that the lumpy, spinning ice ball has a dark crust, like the tartufo dessert (balls of vanilla ice cream coated with chocolate) served in fancy restaurants. Comets aren't so tasty (I think), but they *are* real treats to the eye. Here and there on Halley's nucleus, the probe photographed plumes of gas and dust from geyserlike vents or holes, spraying into space from areas where the Sun was warming the surface. Some crust! And in 2004, NASA's Stardust probe got close-up images of the nucleus of Comet Wild-2. This nucleus seems to bear impact craters and is marked with what may be pinnacles made of ice. Those are the cold facts.

Not all comet nuclei are shaped like Halley's, though. In August 2014, the Rosetta spacecraft reached Comet 67P/Churyumov–Gerasimenko, known as 67P to its friends (like me). Rosetta orbited the comet nucleus while the comet orbited the Sun until the end of the European Space Agency mission in September 2016. Its photographs revealed a nucleus shaped roughly like a dumbbell with two unequal weights. Astronomers referred to the "weights" as two *lobes* of the comet connected by a thinner structure they named the *neck*. Some astronomers stuck their necks out by theorizing that the odd-shaped nucleus was formed by the low-speed collision of two earlier objects.

MORE HIGHLIGHTS FROM ROSETTA'S TRAVEL LOG

Rosetta's photos of 67P distinguished 19 surface areas, which scientists named after ancient Egyptian deities, such as Apis, a sacred bull; Nut, goddess of the sky; and Imhotep, a historical figure who was named a deity after his death. (Wouldn't the family be proud if you were the namesake of a piece of cosmic real estate? It beats the naming rights to a sports stadium, if you ask me.)

The main spacecraft, called the Rosetta orbiter, also deployed a 220-pound (100-kilogram) instrument-bearing lander craft named Philae that unfortunately bounced off the nucleus — twice — before landing at a tilt and, worse, in the shadow of a cliff. Rosetta also gathered a great deal of data on changes to the nucleus (such as the amounts and types of various gases that were streaming off) as the comet orbited closer to the sun, receiving more solar heat.

The coma

As a comet gets closer to the Sun, solar heat vaporizes more of the frozen gas, and it spews out into space, blowing some dust out, too. The gas and dust form a hazy, shining cloud around the nucleus called the *coma* (a term derived from the Latin for "hair," not the common word for an unconscious state). Almost everyone confuses the coma with the head of the comet, but the head, properly speaking, consists of both the coma and the nucleus.

The glow from a comet's coma is partly the light of the Sun, reflected from millions of tiny dust particles, and partly emissions of faint light from atoms and molecules in the coma.

A tale of two tails

The dust and gas in a comet's coma are subject to disturbing forces that can give rise to a comet's tail(s): the dust tail and the plasma tail. (Sometimes when you view a comet, you see just one kind of tail, but when you're lucky, you see both.)

The pressure of sunlight pushes the dust particles in a direction opposite the Sun (see Figure 4-4), producing the comet's *dust tail.* The dust tail shines by the reflected light of the Sun and has these characteristics:

>> A smooth, sometimes gently curved appearance

>> A pale yellow color

The other type of comet tail is a *plasma tail* (also called an ion tail or a gas tail). Some of the gas in the coma becomes *ionized,* or electrically charged, when struck by ultraviolet light from the Sun. In that state, the gases are subject to the pressure of the *solar wind,* an invisible stream of electrons and protons that pours outward into space from the Sun (see Chapter 10). The solar wind pushes the electrified cometary gas out in a direction roughly opposite of the Sun, forming the comet's plasma tail. The plasma tail is like a wind sock at an airport: It shows astronomers who view the comet from a distance which way the solar wind is blowing at the comet's point in space.

In contrast to the dust tail, a comet's plasma tail has the following:

>> A stringy, sometimes twisted, or even broken appearance

>> A blue color

Now and then, a length of plasma tail breaks from the comet and flies off into space. The comet then forms a new plasma tail, much like a lizard that grows a new tail when it loses its first one. The tails of a comet can be millions to hundreds of millions of miles long.

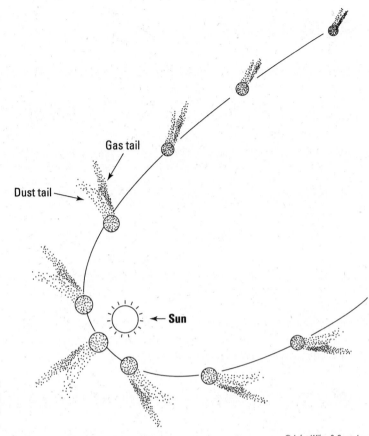

Gas tail

Dust tail

← **Sun**

FIGURE 4-4:
A comet's tail
points away from
the Sun.

When a comet heads inward toward the Sun, its tail or tails stream behind it. When the comet rounds the Sun and heads back toward the outer solar system, the tail still points away from the Sun, so the comet now follows its tail. The comet behaves to the Sun as an old-time courtier did to his emperor: never turning his back on his master. The comet in Figure 4-4 could be going clockwise or counter-clockwise, but either way, the tail always points away from the Sun.

The coma and tails of a comet are just a vanishing act. The gas and dust shed by the nucleus to form the coma and tails are lost to the comet forever — they just blow away. By the time the comet travels far beyond the orbit of Jupiter, where most comets come from, it consists of only a bare nucleus again. And the nucleus is a little smaller, due to the gas and dust that it sheds. The dust the comet loses may someday produce a meteor shower (which I cover earlier in this chapter), if it crosses Earth's orbit.

COMA AGAIN?

The first rule of viewing is, get out of town! Although a comet's nucleus may be just 5 or 10 miles (8 or 16 kilometers) in diameter, the coma that forms around it can reach tens of thousands or even hundreds of thousands of miles across. The gases are expanding, like a puff of smoke from a cigarette. As they thin out, they fade and become less visible. So the size of a comet's coma doesn't depend on just how much stuff the comet sheds; it also depends on the sensitivity of the human eye or the digital camera sensor that you use to observe it. The apparent size of the coma also depends on the darkness of the sky in which you view it. A bright comet looks a lot smaller downtown than out in the country, where the skies are dark.

Halley's comet is a good example of the wasting-away process. Halley's nucleus decreases by at least a meter (39.37 inches, or slightly more than a yard) every 75 to 77 years when it passes near the Sun. The nucleus is only about 10 kilometers (10,000 meters or 6.2 miles) in diameter right now, so Halley's comet will survive only about 1,000 more orbits, or about 75,000 years. Dust shed by the famous comet causes two of the top annual meteor showers, the Eta Aquarids and the Orionids, which I list in Table 4-1.

Waiting for the "comets of the century"

Every few years, a comet is sufficiently bright and in such a good position in the sky that you can easily see it with the naked eye and with small binoculars. I can't tell you when such a comet is coming because the only comets whose returns astronomers can accurately predict in the near future are small ones that don't get very bright. Nearly all bright, exciting comets are discovered rather than predicted.

Halley's Comet is the only bright comet whose visits astronomers can accurately predict, but it doesn't come around often. Its appearance in 1910 was widely heralded, and everyone got a good look. But an even brighter comet came the same year, the Great Comet of 1910, and no astronomer had predicted its arrival. All you can do is keep looking up. Monitor the astronomy magazines and the websites at the end of this section for reports of new comets, and follow the directions to view them. And with luck, you may be the first to spot and report a new comet, in which case the International Astronomical Union will name it after you.

Every five or ten years, a comet comes that is so bright astronomers hail it as "the comet of the century." People have short memories. But stay interested, and you may have a chance to see a fine comet:

>> In 1965, Comet Ikeya-Seki was visible in broad daylight next to the Sun if you held up your thumb to block the bright solar disk. I'll never forget that sight — or my sunburned thumb.

>> In 1976, Comet West was visible to the naked eye even in the night sky over downtown Los Angeles, one of the worst places I know of to see celestial objects. I saw it from there but had a much better view from Arizona.

>> In 1983, Comet IRAS-Iraki-Alcock could be seen by the naked eye, actually moving in the night sky. (Most comets move so slowly across the constellations that you may have to wait an hour or more to notice any change in position.) It looked like a small cloud, blown across the sky when I watched it from a school parking lot in Washington, D.C.

>> In the 1990s, the bright comets Hyakutake and Hale-Bopp appeared out of the blue and were witnessed by millions of people worldwide. Some misguided folks, who listened to too much AM radio, claimed that a UFO was trailing behind the comet. Thirty-nine members of a California cult actually committed mass suicide in the belief that they'd be somehow teleported to the alien spacecraft, which never existed. Please take your comet information only from trusted sources, like those I recommend here.

>> In 2007, Comet McNaught became the brightest comet since Ikeya-Seki in 1965, and it was visible in the daytime for lucky viewers in the Southern Hemisphere.

>> In 2011, Comet Lovejoy, discovered by an amateur astronomer in Australia, came so close to the Sun that it passed though the solar corona, the extremely hot outer atmosphere of the sun (which I describe in Chapter 10). I feared that the corona would be a killjoy for those watching Lovejoy, but the comet survived — just barely.

The next great comet may come at any time. Keep looking, and you may even discover it!

TIP

Plenty of websites offer information on currently visible comets and photographs of them from amateur and professional astronomers. Most of the time, the current comets are too dim for any but advanced amateur telescopes. Check one or two of these good resources regularly to make sure you have the latest word:

>> The Comet Chasing page gives finder charts and detailed information on the visibility of comets each month for observers at various latitudes around the world, from 55° north to 30° south, at cometchasing.skyhound.com. A *finder chart* is a detailed map of the stars in a region of the sky where a comet is expected to pass. It usually marks the predicted path of the comet, with tick marks that indicate which dates the comet will be visible at various places along the path.

>> Southern Sky Watch in Australia provides information on comets and other sky objects visible from Down Under; check it out at www.users. on.net/~reynella/skywatch/ssky.htm.

>> The Heavens-Above website provides star charts for current comets; visit www.heavens-above.com. Thank heavens for them.

Hunting for the great comet

Finding a comet isn't difficult, but finding your first one can take years and years. The famous comet hunter David Levy scanned the sky systematically for nine years before he found his first comet. Since then, he's found over 20 more.

The best telescope to use for comet searching is a *short focus* or *fast* telescope, meaning one whose catalogue specifications include a low f-number (like the f-number of a camera lens) — f/5.6 or, better yet, f/4. And you need to use a low-power eyepiece, such as 20x to 30x (see Chapter 3). The whole idea of the low f/number and the low magnification is to view as large an area of the sky as possible with your telescope. (It's called wide field observing.) The bright comets that you may be able to discover are few and far between, so you've got to look far and wide to catch them.

TIP

A quality, modestly priced telescope to start your comet hunt is the Orion ShortTube 80-A Equatorial Refractor, with an 80-millimeter (3.1-inch) objective lens. Its f/5.0 focal ratio and wide field eyepieces are just what you need for comet searches. This instrument lists for about $200 on the Orion Telescopes and Binoculars website, www.telescope.com. However, you also need a good stand for the telescope. The Orion EQ-1 Equatorial Telescope Mount, about $130, is suitable. (I explain terms like refractor, objective lens, and much more in Chapter 3.) Celestron offers a worthy, inexpensive telescope of another type, the Cometron 114AZ, for comet hunting. It's a 114-millimeter (4.5-inch) Newtonian reflector (see Chapter 3) that comes with an alt-azimuth mount and tripod. The cost of the whole shebang is about $180. See www.celestron.com.

You can search for unknown comets in two ways: the easy way and the systematic way. Read on to discover both techniques and for info on reporting a comet.

Locating comets the easy way

The easy way to search for comets is to make no extra effort at all. Just be on the lookout for fuzzy patches when you stare through your binoculars or telescope at stars or other objects in the night sky. Scan the sky for a fuzzy spot (as opposed to stars, which are sharp points of light if your binoculars are in focus). If you pinpoint a fuzzy area, check your star atlas to see whether anything at that location is *supposed* to look fuzzy, such as a nebula or a galaxy. If you find nothing like that on the atlas, you may have found a comet — but before you get too excited, wait a few hours and see whether the possible comet moves against the pattern of adjacent stars. If the Sun rises or clouds move in the way and block your view, look again on the next night. If the object is indeed a comet, you'll notice that its position has changed with respect to the stars. And if the fuzz is bright enough, you may be able to spot a tail, which is a dead giveaway that you've found a comet.

Locating comets the systematic way

The systematic way to search for comets is based on the precept that you can find them most easily where they're brightest and where the sky is at its darkest. Comets closest to the Sun are the brightest, but the sky is darkest in directions far from the direction to the Sun.

As a compromise between as far from the Sun and as close to the Sun as possible, look for comets in the east before dawn over the part of the sky that's both

>> At least 40° from the Sun (when the latter is below the horizon)

>> No more than 90° from the Sun

Remember that there are 360° all the way around the horizon, so 90° is one-quarter of the way around the sky.

A desktop planetarium program can help you map out the regions of the constellations that fit this bill for any given night of the year (see Chapter 2 for more about these programs). And, of course, you can look for comets in the west at dusk by following the same two rules about distance from the Sun. In my experience, the first few "comets" that you discover will be the contrails from jet airplanes, which catch the rays of the Sun at their high altitude, even though the Sun has set at your location on the ground.

TIP

Start at one corner of the sky area that you plan to check and slowly sweep the telescope across the area. Move the telescope slightly up or down and scan the next strip of sky in your search area. You can make every scan from left to right, or you can scan back and forth boustrophedonically (a term from classical times that refers to plowing a field with oxen; the oxen plow the first furrow in one direction and then come back across the field, plowing in the opposite direction).

You'll have an easier time impressing your friends by telling them of your boustrophedonic comet search project than by actually discovering a comet. It will give your ego a boost (unless your friends decide that you're plowed).

PLAYING THE SPACE OBJECT NAME GAME

If you discover a comet, the International Astronomical Union will name it after you and possibly also after the next one or two people who independently report it.

If you discover a meteor, you don't have time to name it before it disappears. You can try shouting "John," but the name won't catch on, and you may attract undue attention. The only meteors that get named are the spectacular ones seen by thousands of people. They get names such as the "Great Daylight Fireball of August 10, 1972," but no official procedure governs how meteors get named.

If you discover a meteorite, it's named for the town or other local area where you pick it up. The meteorite belongs to the owner of the land where you find it, and if you find it on U.S. government land, such as a national park or forest, it goes to the Smithsonian Institution.

If you discover an asteroid, you can recommend a name for the asteroid, but it can't be your name. On the other hand, someone else who finds an asteroid can nominate you for the honor. (See Chapter 7 for more about asteroids.)

Reporting a comet

When you discover a comet, follow the directions on the website of the International Astronomical Union's Central Bureau for Astronomical Telegrams (which doesn't use telegrams anymore) and report it by email. The site is www.cbat.eps.harvard.edu.

The bureau doesn't appreciate false alarms, so try to get a stargazing friend to check out your discovery before you spread the word. If the find checks out, you — as the amateur discoverer of a comet — may be eligible for a cash share (or all) of the Edgar Wilson Award, which is described on the Central Bureau website.

But even if you never discover a comet, and most astronomers never do, you can enjoy comets that others find, and tell tales of them to your friends.

Artificial Satellites: Enduring a Love–Hate Relationship

An artificial satellite is something that people build and launch into space, where it orbits Earth or another celestial body. The Earth-orbiting artificial satellites show us the weather, monitor El Niño, relay network television programs, and stand guard against intercontinental missile launches by hostile powers. And they can also be used for astronomy.

The Hubble Space Telescope is an artificial satellite, and astronomers love it. It gives unparalleled views of the stars and distant galaxies and lets you view the universe in ultraviolet and infrared light that's otherwise blocked by the thick layers of Earth's atmosphere. (NASA plans to keep the Hubble working for a while longer, but some time in the 2020s, at the latest, it will be deorbited and will fall into the sea.)

Artificial satellites can catch the rays of the setting Sun or even the Sun that has already set for observers at ground level. Capturing the light of the Sun, they represent points of light that may move across the part of the sky where an astronomer is making a photograph of faint stars. Astronomers don't appreciate this interference. Worse yet, some artificial satellites broadcast at radio frequencies that interfere with the "big dish" and other radio antennas that astronomers use to receive the natural radio emanations from space. The celestial radio waves may have traveled for 5 billion years from a quasar, or they may have taken 5,000 years to reach us from another solar system in the Milky Way, possibly bearing a greeting from benevolent aliens who want to send us the cure for cancer. But just as the radio waves arrive, a blaring tone and strident modulations from a satellite passing over the observatory interfere with our reception. We may never know what the news is from the planet of the aliens.

So astronomers love satellites when they do something good for us and hate them when they interfere with our observations. But to make the best of a bad thing, amateur astronomers have become enthusiastic viewers and photographers of artificial satellites passing overhead.

Skywatching for artificial satellites

Hundreds of operating satellites are orbiting Earth, along with thousands of pieces of orbiting space junk — nonfunctional satellites, upper stages from satellite launch rockets, pieces of broken and even exploded satellites, and tiny paint flakes from satellites and rockets.

You may be able to glimpse the reflected light from any of the larger satellites and space junk, and powerful defense radar can track even very small pieces.

The best way to begin observing artificial satellites is to look for the big ones — such as NASA's International Space Station or the Hubble Space Telescope — and the bright, flashing ones (the dozens of Iridium communication satellites).

Looking for a big or bright artificial satellite can be reassuring to the beginning astronomer. Predictions of comets and meteor showers are sometimes mistaken, the comets usually seem fainter than you expect, and usually you see fewer meteors than advertised. But artificial satellite viewing forecasts are usually right on. You can amaze your friends by taking them outside on a clear early evening, glancing at your watch, and saying "Ho hum, the International Space Station should be coming over about there (point in the right direction as you say this) in just a minute or two." And it will!

Want to know what to watch for? I've got you covered. Here are some characteristics you can pinpoint for both large and bright satellites:

» A big satellite such as the Hubble Space Telescope or the International Space Station generally appears in the evening as a point of light, moving steadily and noticeably from west to east in the western half of the sky. It moves much too slowly for you to mistake it for a meteor, and it moves much too fast for a comet. You can see it easily with the naked eye, so it can't be an asteroid — and, anyway, it moves much faster than an asteroid.

Sometimes you may confuse a high-altitude jet plane with a satellite. But take a look through your binoculars. If the object in view is an airplane, you should be able to distinguish running lights or even the silhouette of the plane against the dim illumination of the night sky. And when your location is quiet, you may be able to hear the plane. You can't hear a satellite.

» An Iridium satellite is a wholly different viewing situation: It usually appears as a moving streak of light that gets remarkably bright and then fades after several seconds. It moves much more slowly than a meteor. And an Iridium flare or flash is often brighter than Venus, second in brilliance only to the Moon in the night sky. The Sun, located below your horizon, reflects off one of the door-size, flat, aluminum antennas on the satellite to cause the flash of light. At star parties, people cheer when they spot an Iridium flare, just like when folks see a fireball. You can even see some Iridium flares in daylight.

And consider this: More than 60 Iridium satellites are in orbit. They interfere with astronomy, and professional astronomers want them to disappear, but until now at least the satellites have had a "flare" for entertaining us. A new generation of the satellites, called Iridium NEXT, are being launched (the first ten went up into space in January 2017). The NEXT satellites may be next to useless for amateur flare watchers because the design of the antennas has changed so that bright reflections from them are unlikely. The good news is that retiring all the original Iridiums will take a while, so if you start looking soon you may be able to catch some impressive flares before they're just history.

Finding satellite viewing predictions

TIP

Some newspapers and television weather persons give daily or occasional forecasts for viewing satellites from the local area. You can get more detailed information whenever you want it by consulting these websites:

>> For the International Space Station and the Hubble Space Telescope, *Sky & Telescope* offers a Satellite Tracker. Go to its site, www.skyandtelescope.com and register (it's free) as a site user. Then click on the Observing menu, select Interactive Tools, open the Satellite Tracker descriptive text, and hunt within it for the Satellite Tracker link. Click the link, which opens a window where you enter the names of your city and country and get viewing predictions. It sounds complicated, but it sure beats trying to calculate satellite passes yourself.

>> For Iridium communication satellites, convenient forecasts are available from the Heavens-Above website at www.heavens-above.com. This site also provides predictions for daylight viewing of Iridium flares. Note that when you open the Iridium flares page, you must enter your geographic location to get the correct satellite predictions; otherwise, you'll get predictions for a default location. Don't fall for them.

>> Heavens-Above is also another good source place for Hubble Space Telescope viewing info (www.heavens-above.com). Click the "HST" link in the "Satellites" section to get a viewing schedule after you enter your location.

You can also use a smartphone app to locate a bright satellite:

>> If you have an iPhone, use Go Sat Watch or Sky Safari.

>> For Android phones, Sky Safari works too.

After you succeed in viewing some bright artificial satellites, you can try photographing them. Follow the directions in the sidebar "Photographing meteors and meteor showers" earlier in this chapter. I don't recommend using point-and-shoot digital cameras for photographing meteors or most satellites. However, the International Space Station may be bright enough to catch with that kind of camera.

2

Going Once Around the Solar System

IN THIS PART . . .

Find out more about our home planet, Earth, and its Moon.

Meet Earth's nearest neighbors: Mercury, Venus, and Mars.

Take a ride through the asteroid belt and find out whether Earth is in any danger of being struck by one of these mighty rocks.

See why so many people are fascinated by Jupiter and Saturn, two giant balls of gas.

Get details on Uranus and Neptune, the most distant known planets in our solar system, and the dwarf planet Pluto, which lies beyond them.

Chapter **5**

A Matched Pair: Earth and Its Moon

People often think of planets as objects up in the sky, like Jupiter and Mars. The ancient Greeks — and people for centuries thereafter — made a distinction between Earth, which they regarded as the center of the universe, and the planets. They thought of the planets as little lights in the sky that revolved around Earth.

Today we know better. Earth is a planet, too, not the center of the universe. It isn't even the center of our solar system; the Sun owns that title. The Moon orbits around Earth, along with hundreds of artificial satellites (see Chapter 4), and that's about it. Joining Earth in orbit around the Sun are seven other planets in the solar system, Pluto and several other objects called "dwarf planets," a large number of other moons, a belt of asteroids, millions of comets, and more. Nevertheless, as far as we know, life exists in our solar system only on Earth.

Earth has fallen from its exalted place in human thought as the center of the universe to its true, but still significant, status: our home planet. And no other place in the solar system is quite like home.

Earth is what astronomers call a *terrestrial* planet — a kind of circular definition because *terrestrial* means "earthly." But the scientific meaning is a planet made of rock that orbits the Sun. The four planets closest to the Sun are our solar system's terrestrial planets: Mercury, Venus, Earth, and Mars, in order of distance from the Sun.

Some people regard the Earth–Moon system as a double world. For aliens seeking to visit us, that distinction may help: "Just head for the yellow-white star in Sector 49,832 of the Orion Arm, in the Milky Way, and home in on the third rock from that Sun; it's a double world and easy to spot."

Putting Earth under the Astronomical Microscope

Earth is unique among the known planets. In the following sections, I tell why, briefly summarizing some of its main characteristics and how they play into astronomical topics like time and the seasons. And in case you forgot what it looks like, you can check out a nice NASA photo in the color section, which shows Earth and the Moon together.

One of a kind: Earth's unique characteristics

What's so special about Earth? For starters, we inhabit the only planet in the solar system with all these known characteristics:

>> **Liquid water at the surface:** Earth has lakes, rivers, and oceans of water, unlike any other known planet. Unfortunately, it has tsunamis and hurricanes, too. The oceans cover 70 percent of the surface of Earth.

>> **Plentiful amounts of oxygen in the air:** The air on Earth contains 21 percent oxygen; no other planet has more than a trace of oxygen in its present atmosphere, as far as we know. (Most of Earth's atmosphere, about 78 percent, is nitrogen.)

>> **Plate tectonics (also known as *continental drift*):** Earth's crust is composed of huge moving plates of rock, where plates collide, earthquakes occur, and new mountains rise. New crust emerges at the midocean ridges, deep beneath the sea, causing the seafloor to spread. (To find out about an interesting seafloor property, see the sidebar "Earth's seafloor and its magnetic properties" later in this chapter.)

>> **Active volcanoes:** Hot molten rock, welling up from deep beneath the surface, forms huge volcanic landforms such as the Hawaiian Islands. Volcanoes erupt somewhere on Earth every day.

>> **Life, intelligent or otherwise:** I'll let you be the judge on intelligence, but from one-celled amoebas, bacteria, and viruses to flowers and trees, fish and fowl, and insects and mammals, Earth has life in abundance.

Researchers are investigating tantalizing indications that Mars and Venus may once have shared some of these traits with Earth (see Chapter 6). But as far as we know, they don't have life now, and we don't have proof that they ever did.

Scientists believe that the presence of liquid water on the surface of Earth is one of the main reasons life flourishes here. You can easily imagine advanced life forms on other worlds — you see them on television and in the movies. But the images you see are all imaginary. Scientists don't have convincing evidence for any life, past or present, anywhere but on Earth. (For more about the possibility of life elsewhere in space and the search for extraterrestrial intelligence, see Chapter 14.)

ENJOYING THE NORTHERN LIGHTS

The aurora is one of the most beautiful sights of the night sky and, for many people, a rare one. Depending on whether you live in the Northern Hemisphere or the Southern Hemisphere, you can see the *aurora borealis* (northern lights) or the *aurora australis* (southern lights), respectively.

Auroras appear when streams of electrons from Earth's magnetosphere rain down on the atmosphere below, stimulating oxygen and other atoms to shine. The eerie glow in the night sky may remain stationary for minutes to hours or constantly change (making it hard for a novice observer to identify). It can shimmer, pulsate, or flash around the sky. The aurora appears in many forms; here are a few of the most common:

- **Glow:** The simplest aurora. It looks like thin clouds are reflecting moonlight or city lights. But you don't see clouds — just the eerie light of an aurora.

- **Arc:** Shaped like a rainbow but with no sunlight to make one. A steady or pulsating green arc is the most common type of arc, but sometimes dim red arcs appear.

- **Curtain:** Also called drapery. This spectacular aurora resembles a billowing curtain at a theater, but nature is the star of the show.

- **Rays:** One or more long, thin bright lines in the sky, appearing like faint beams from the heavens.

- **Corona:** High overhead, a crown in the sky, with rays emanating in every direction.

(continued)

(continued)

Auroras occur constantly in two geographical bands around Earth at high northern and southern latitudes. Folks who live beneath these two *auroral ovals* see auroras every night. But you may encounter big exceptions: When a great disturbance in the solar wind (see Chapter 10) strikes Earth's magnetosphere, the ovals move toward the equator. People in the *auroral zones* (the lands beneath the ovals) may miss their aurora, but skygazers toward the equator who rarely see them are treated to a great show. The most likely times to see bright auroras outside the auroral zones are the first few years after the peak of the sunspot cycle, so keep your eyes open for auroras in 2024 and the next few years.

Tip: If you do spot an aurora, please help scientists by making a simple report. Just surf to the citizen science website aurorasaurus.org. Then click "Yes" where it asks "Did you see the aurora?" Next, fill out the simple form that pops up, and you've done your part.

If you don't want to wait for aurora to come to you, visit a high-latitude location, where you're near the northern auroral oval and can see aurora on almost any clear, dark night. Famous aurora-viewing places include

- **Fairbanks, Alaska** (fairbanks-alaska.com/northern-lights-alaska.htm)
- **Yellowknife, Canada,** capital of the Northwest Territories, where you can relax in the Aurora Village's teepees "heated by a wood stove and stocked with hot beverages" and enjoy the sky show (auroravillage.com)
- **Tromsø, Norway,** where you can visit "northern lights camps" and go snowmobiling or reindeer sledding while aurora watching (Norway-lights.com/#tromso)

The good news about these northern locales is that you'll have great auroral views (weather permitting). The bad news is that the best time to see aurora in these cold places is from December to March, the coldest part of the year. At other times, the nights are shorter, so viewing prospects are poorer.

If you live in the Northern Hemisphere and want to look for aurora, check the daily aurora borealis forecast at the University of Alaska Geophysical Institute website at www.gi.alaska.edu/AuroraForecast/NorthAmerica. Better yet, get the forecast on your smartphone; just download the My Aurora Forecast app (free for iPhone and iPad) or My Aurora Forecast Pro (about $2.00).

If the aurora isn't visible at your location, head to the Aurora Borealis Notifications site at auroranotify.com and click on "Webcams, Links, & Apps" Then you see a menu of live camera stations in Alaska, the northern continental United States, Canada, Sweden, Norway, Finland, and more. There are even links to cameras viewing the aurora australis in Tasmania and Antarctica. User beware: At any given time, some cameras may be out of operation, and some others aren't sending aurora photos because it's daytime at their sites or the sky is clouded over.

Spheres of influence: Earth's distinct regions

Figure 5-1 shows Earth as seen from space. Earth's patterns of land, sea, and clouds are clearly visible.

FIGURE 5-1:
Earth photographed by the Deep Space Climate Observatory.

Scientists classify the regions of Earth into these categories:

>> The *lithosphere:* The rocky regions of our planet

>> The *hydrosphere:* The water in the oceans, lakes, and elsewhere on Earth

>> The *cryosphere:* The frozen regions — notably, the Antarctic and Greenland ice caps

>> The *atmosphere:* The air from ground level up to hundreds of miles

>> The *biosphere:* All living things on Earth — on land, in the air and water, and underground

So you're part of the biosphere that lives on the lithosphere, drinks from the hydrosphere, and breathes the atmosphere. (You can also visit the cryosphere.) I don't know anywhere else in space you can do all that.

In addition to the regions I describe in the previous list, one more important part of our planet is the *magnetosphere,* which plays a vital role in protecting Earth from many of the dangerous emanations from the Sun (see Chapter 10). Inside the magnetosphere are regions where electrically charged particles — mostly electrons and protons — bounce back and forth above Earth, trapped in its magnetic field. These regions are sometimes called Earth's radiation belts (or the Van Allen radiation belts). (The Van Allen radiation belts are named for James Van Allen, a U.S. physicist who discovered them with America's first artificial satellite, Explorer 1.)

Occasionally, some of the electrons escape and rain down on Earth's atmosphere below, striking atoms and molecules and making them glow. That glow is the aurora (see the earlier sidebar "Enjoying the northern lights" for more about viewing auroras).

The solid surface of Earth — the part you stand on — is the crust. Beneath the crust are the mantle and the core. The core is largely iron and nickel and is very hot, reaching about 10,800°F (about 6,000°C) at the center. The core is layered, too: The outer core is in a molten state, but the inner core is solid.

The extremely high pressure of the overlying layers makes the hot iron in the inner core solidify. As Earth cools over millions of years in the future, the solid part at the center increases in size at the expense of the surrounding molten core, like an ice cube growing as the surrounding liquid cools.

Earth's core is far below our digging range, but it produces an effect that anyone can observe at the surface. Moving streams of molten iron in the outer core generate a magnetic field that reaches out through the whole planet and far into space, called the *geomagnetic field.*

The geomagnetic field

>> Makes the needle of a compass point toward north (or south)

>> Provides an invisible guidance system for homing pigeons, some migratory birds, turtles, salmon, various ants and bees (among other bugs), and even some ocean-dwelling bacteria

>> Forms the magnetosphere far above Earth

>> Shields Earth from incoming electrically charged particles from space, such as the solar wind and many cosmic rays (high-speed, high-energy particles that come from explosions on the Sun and from distant points in space)

The geomagnetic field is a global planetary magnetic field, meaning that it extends above all parts of Earth and is continuously being generated. Mars, Venus, and our Moon all lack a global magnetic field like Earth's, and this key difference gives scientists information about the cores of those objects. For more on the lunar core, see the section "Quite an impact: Considering a theory about the Moon's origin" later in this chapter.

EARTH'S SEAFLOOR AND ITS MAGNETIC PROPERTIES

According to geophysical surveys, patterns of magnetized rock exist in the seafloor on either side of midocean ridges. The rock became magnetized as it cooled from the molten state, trapping and "freezing in" some of Earth's magnetic field that pervaded it as the rock solidified. So the seafloor rock resembles a magnet, with a magnetic field that has strength and direction. After the rock solidified, its magnetic field could change no longer, and it became a fossil magnetic field. It's like a fossil dinosaur that remains forever in the shape it had at death.

The patterns discovered near the midocean ridges consist of stripes of magnetized rock, hundreds of miles long, that parallel the ridges and alternate in polarity. One stripe has a north magnetic polarity, like the end of a bar magnet that attracts a north-seeking compass needle; the next stripe has the opposite polarity; and so on.

The alternating stripes of oppositely magnetized rock are the result of the new rock emerging from the midocean ridges, cooling and magnetizing, and moving away from the ridges as even newer rock pushes it along. The oppositely magnetized stripes show that the geomagnetic field itself reverses direction from time to time, like a bar magnet that you turn 180° at intervals — except that the intervals for the geomagnetic field range from about a thousand to a million years.

An unknown process causes the geomagnetic field, generated deep in Earth's core, to reverse every so often. That effect is preserved in the fossil magnetic fields of the rock at the seafloor and in rock on the continents that previously lay beneath the sea.

Why mention all this seafloor stuff in a book on astronomy? Because this unique property of Earth may correspond to a phenomenon discovered on Mars. As scientists consider the evidence gathered on the various terrestrial planets, including Earth, we find similarities and differences that help us understand them better. Such research is called *comparative planetology*, and I cover it in more detail in the descriptions of Mars and Venus in Chapter 6.

Examining Earth's Time, Seasons, and Age

The rotation of Earth was the original basis of our system of measuring time, and we now know that the orbital motion of Earth and the tilt of its axis produce the seasons. Our seasonal ring-around-the-rosy (or yellowy, in this case) sun dance has been going on for a long time; Earth is about 4.6 billion years old.

Orbiting for all time

Nowadays, scientists have atomic clocks that measure time with great precision. But originally and until modern times, our system of time was based on the rotation of Earth.

Knowing how time flies

Earth rotates once on its axis in 24 hours. It turns from west to east (which is counterclockwise as seen from space above the North Pole). The length of the day, 24 hours, is the average time it takes for the Sun to rise and set and rise again. This process is called *mean solar time* and is equivalent to the standard time on your watch.

The length of the day, therefore, is 24 hours of mean solar time. And a year consists of approximately 365 days, the time that it takes Earth to make one complete orbit around the Sun.

Because Earth moves around the Sun, the time when you see the Sun rise depends on both the rotation of Earth and Earth's orbital motion.

Earth turns once in 23 hours, 56 minutes, and 4 seconds with respect to the stars. That amount of time is called the *sidereal day*. (*Sidereal* means "pertaining to the stars.") Notice that the difference between 24 hours and 23 hours, 56 minutes, and 4 seconds is 3 minutes and 56 seconds, which is just about $\frac{1}{365}$ of a day. The difference is no coincidence: It happens because, during a day, Earth moves through $\frac{1}{365}$ of its orbit around the Sun.

TECHNICAL STUFF

Astronomers used to depend on special clocks called sidereal clocks, which measured sidereal time by registering 24 sidereal hours during an interval of 23 hours, 56 minutes, and 4 seconds of mean solar time. The sidereal hours, minutes, and seconds are all slightly shorter than the corresponding units of solar time. Using sidereal clocks enabled astronomers to keep track of the stars in order to point telescopes correctly. But we don't need to do that anymore. Computer programs that point telescopes or that picture the sky on a desktop planetarium or smartphone, as I describe in Chapter 2, do the mathematics for you, so you can simply use the standard time at your location to figure out where different stars and constellations appear in the sky.

On the other hand, astronomers still adhere to the custom of reporting astronomical observations on a common system called *Universal Time (UT)* or *Greenwich Mean Time*. UT is simply the standard time at Greenwich, England. If you live in North America, the standard time at your location is always earlier than the time in Greenwich. For example, in New York City, the Sun rises about 5 hours after it rises in Greenwich. When the clock strikes 6 a.m. in Greenwich, clocks in New York are turning to 1 a.m.

A more precisely defined time, *Coordinated Universal Time*, or *UTC*, which is identical to UT for all practical purposes, is the official international standard.

Looking out for leap seconds

Earth takes 365¼ days to go around the Sun, but a normal calendar year is just 365 days. That's why we add a day, February 29, every fourth year. We call that fourth year *leap year*, and the added day helps to keep Earth and the calendar in sync.

There's also a sync problem between the rotation of Earth and the length of a calendar day: Every now and then, Earth turns a wee bit slower than usual, perhaps due to a large weather system such as El Niño. These discrepancies accumulate and, before you know it, Earth is turning is out of whack with time kept by ultraprecise atomic clocks. On such occasions, international authorities declare a *leap second,* which is added to the UTC. When that little adjustment is made, engineers worldwide hold their breath and cross their fingers, hoping that it doesn't mess up GPS, air traffic control, and other critical, time-dependent systems.

That's why some nations would like to see leap seconds abolished so that we would just rely on atomic clocks. If that were done, our time system would no longer be synched to Earth, and eventually noon on the clock might happen when the Sun was just rising, or even when it was nighttime. A leap second was added on December 31, 2016. Depending on where the decision on abolishing leap seconds lands, this date may have been the last small leap for mankind's clocks. Only time will tell.

Finding the right time

TIP

In the United States, the U.S. Naval Observatory (USNO) in Washington, D.C., is in charge of time. You can get the UTC any time you want it from the USNO home page at www.usno.navy.mil/USNO.

TIP

To determine the standard time zone that applies at just about any other place in the world and to convert it to Universal Time, consult the Time Zone Map at www.timeanddate.com/time/map.

Generally, daylight saving time (called British Summer Time in the United Kingdom) is an hour later than standard time at the same geographic location. But not all places observe daylight saving time. For example, Arizona, which gets plenty of sunshine year-round, never goes on daylight saving time.

Tilting toward the seasons

Teaching students about the cause of the seasons is about the most frustrating task of any astronomy professor. No matter how carefully the professor explains that the seasons have nothing to do with how far we are from the Sun, many students don't grasp it. Even surveys taken at Harvard University graduation show that bright college graduates think that summer occurs when Earth is closest to the Sun and winter occurs when Earth is farthest from the Sun.

Students forget that when summer comes in the Northern Hemisphere, the South experiences winter. And when Australians are surfing in the summer, people in the United States are wearing their winter coats. But Australia and the United States are on the same planet. Earth can't be farthest from the Sun and closest to the Sun at the same time. Earth is a planet, not a magician.

The actual cause of the seasons is the tilt of Earth's axis (see Figure 5-2). The *axis*, the line through the North and South poles, isn't perpendicular to the plane of Earth's orbit around the Sun. Actually, the axis is tilted by $23\frac{1}{2}°$ from the perpendicular to the orbital plane. The axis points north to a place among the stars — in fact, near the North Star (at least, in the short term; the axis slowly changes its pointing direction, so the North Star in one era will no longer be the North Star in the far future).

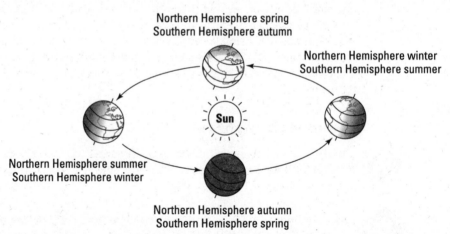

FIGURE 5-2:
The tilt of Earth's axis determines the seasons.

© John Wiley & Sons, Inc.

At present, the North Star, also called Polaris, is the star Alpha Ursae Minoris, located in the Little Dipper asterism of the Little Bear constellation, Ursa Minor. If you get lost at night and want to "bear" north, set your sights on the Little Dipper. (See Chapter 3 for more about finding Polaris.)

The axis of Earth points "up" through the North Pole and "down" through the South Pole. When Earth is on one side of its orbit, the axis pointing up also points roughly toward the Sun so the Sun looms high in the sky at noon in the Northern Hemisphere. Six months later, the axis points up and roughly away from the Sun.

Actually, the axis always points in the same direction in space, but Earth has now moved to the opposite side of the Sun.

Summer occurs in the Northern Hemisphere when the axis pointing up through the North Pole points roughly at the Sun. When that happens, the Sun at noon is higher in the sky than at other seasons of the year, so it shines more directly on the Northern Hemisphere and provides more heat. At the same time, the axis pointing down through the South Pole points away from the Sun, so it shines lower in the sky at noon than at any other season of the year, creating less direct sunlight — and, thus, winter in Australia.

We enjoy more hours of sunlight in the summer because the Sun is higher in the sky. It takes longer for the Sun to rise to that height, and it takes longer to set.

As we orbit the Sun, it seems to move through the sky, following a circle called the *ecliptic*, which I mention in Chapter 3. The ecliptic is tilted with respect to the equator by exactly the same angle as Earth's tilt on its axis: $23\frac{1}{2}°$. Here are some key events in the Sun's annual journey around the ecliptic, as experienced in the Northern Hemisphere:

>> **Vernal equinox:** On the first day of spring, the Sun crosses from "below" (south) the equator to "above" (north).

>> **Summer solstice:** The Sun reaches the farthest point north on the ecliptic.

>> **Autumnal equinox:** The Sun crosses the equator going back down south, and fall begins.

>> **Winter solstice:** The Sun gets as far south as possible on the ecliptic.

In the Northern Hemisphere, the summer solstice is the day with the most hours of sunlight during the year because the Sun attains its highest position in the sky — it takes the longest time to reach that height and come back down to the horizon again. By the same token, the winter solstice in the Northern Hemisphere is the day with the shortest amount of daylight during the year.

And that's the long and the short of time and seasons.

Estimating Earth's age

Measuring radioactivity is the only accurate way we have to date very old things on Earth or in the solar system. Some elements, such as uranium, have unstable forms called *radioactive isotopes.* A radioactive isotope turns into another isotope of the same element, or into a different element, at a rate determined by the *half-life* of the radioactive substance. If the half-life is 1 million years, for example, half of the radioactive isotope that was originally present will have turned into another substance (called the *daughter isotope)* by the time 1 million years has elapsed, leaving half still radioactive. And half of the remaining half turns into daughter isotope atoms in another million years. So after 2 million years, only 25 percent of the original radioactive isotope atoms still exists. After 3 million years, only $12\frac{1}{2}$ percent remains. And so on.

When the original radioactive isotope atoms, called the *parent atoms,* and the daughter atoms are trapped together in a piece of rock or metal, such as a meteorite, scientists can count the atoms' respective numbers to determine how old the rock is in a process called *radioactive dating.*

Scientists have used radioactive dating to determine that the oldest rocks on Earth are about 4 billion years old. However, Earth is undoubtedly older than that. Erosion, mountain building, and *volcanism* (the eruption of molten rock from within Earth, including the formation of new volcanoes) constantly destroy the rocks at the surface, so the original surface rocks of Earth are long gone.

Meteorites, however, yield radioactive dates as old as 4.6 billion years. Meteorites are considered debris from asteroids, and asteroids are thought to be debris from the very early solar system, when the planets first formed (I cover meteorites in Chapter 4; see Chapter 7 for more about asteroids).

So scientists think that Earth and other planets are about 4.6 billion years old. Earth's Moon, however, is a little younger, as I explain in the next section.

Making Sense of the Moon

The Moon is 2,160 miles (3,476 kilometers) in diameter, slightly more than a quarter of the diameter of Earth. The Moon has no meaningful atmosphere — just a trace of hydrogen, helium, neon, and argon atoms, along with other traces in even lesser quantities. It's all, or mostly all, made of solid rock (see Figure 5-3); some experts think it may have a small molten iron core. (Not a speck of cheese in sight.) Its mass is only $\frac{1}{81}$ the mass of Earth, and its density is about 3.3 times the density of water, which is noticeably less than the density of Earth (5.5 times the density of water).

The following sections give you the lowdown on the Moon's phases, lunar eclipses, and its geology (including handy tips for viewing a variety of lunar features). I also share a theory about the Moon's origin.

FIGURE 5-3:
The Moon is
made of rocks
and rilles, craters,
and dried lava
plains.

Courtesy of NASA

Get ready to howl: Identifying phases of the Moon

Except during a lunar eclipse (see the next section), half the Moon is always in sunlight and half is always in night. But contrary to popular belief, these light and dark hemispheres don't correspond to the lunar near side and the lunar far side. Those sides are the hemispheres that point toward and away from Earth, which are always the same. The lunar halves in sunlight and in night are the hemispheres that face toward and away from the Sun. And they always change as the Moon moves around Earth (see Figure 5-4).

New Moon is the beginning of the monthly lunar cycle, or *lunation*. At this time, the near side faces away from the Sun, making it the dark side. A few hours or days later, the Moon is a new crescent, or *waxing crescent*, meaning a crescent Moon whose bright area is getting larger. This phase happens as the Moon moves away from the Sun–Earth line while orbiting Earth. Fully half of the Moon is always lit up, facing the Sun, but during a crescent Moon, we can't see most of this illuminated area that faces away from Earth.

As the Moon moves around its orbit, it reaches a point where the Earth–Moon line is at right angles to the Earth–Sun line. At this stage, we see a *half Moon*, which astronomers call a *quarter Moon*.

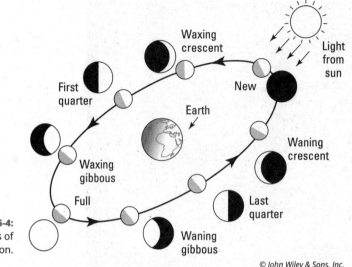

FIGURE 5-4:
The phases of
the Moon.

How can a half equal a quarter? It can't if you're trying to make change, but an astronomer can easily make it work. Half of the lunar near side — the part facing Earth — is lit up, so people call it a half Moon. But the illuminated portion of the Moon that we see is only half the bright hemisphere that faces the Sun, and half of a half is a quarter. Bet your friends that a quarter can be a half. You'll win, and you can pocket the change.

When the illuminated part of the Moon that we can see grows larger than the quarter (half) Moon but is still smaller than the full Moon, astronomers call it a *waxing gibbous Moon.*

When the Moon is on the far side of its orbit, opposite the Sun in the sky, the lunar hemisphere that faces Earth is fully lit, creating a *full Moon.* As the Moon continues around its orbit, the illuminated portion gets smaller and the Moon becomes gibbous again, less than full and more than a quarter Moon (a *waning gibbous Moon*). Soon the Moon appears as a quarter Moon again, called *last quarter.* As the Moon nears the line between Earth and the Sun, it becomes a *waning crescent Moon.* Soon it becomes a new Moon again, and the cycle of phases starts over.

The period of time over which the lunar phases change from new Moon to the next new Moon is the *synodic month,* which averages 29 days, 12 hours, and 44 minutes.

I hope you won't be disappointed when I tell you that the Moon is on the way out. The Moon is slowly moving away from Earth while still orbiting around it. Every year, the Moon recedes by about 1½ inches, and the time it takes for one orbit around Earth gets very slightly longer.

TIP

For reliable data on the phases of the Moon, visit the site of Her Majesty's Nautical Almanac Office at `astro.ukho.gov.uk/nao/online`, scroll down to "Astronomical and Calendarial Data Sheets," click on the current year, and you can, view, download, or print an Astronomical Information Sheet with the exact dates and times of the phases of the Moon throughout the year and other interesting data.

People often ask why an eclipse of the Sun doesn't occur every month at new Moon. The reason is that Earth, the Moon, and the Sun usually aren't all exactly on a line at a new Moon. When they are all in line at new Moon, an eclipse of the Sun results (see Chapter 10). When the three bodies are all on a line at a full Moon, we witness an eclipse of the Moon.

FEELING BLUE? CATCH A SUPERMOON

You may see a supermoon perhaps only once in a blue moon, but you can count on observing the Moon illusion as much as you like. Here's what these terms mean:

- A *supermoon* occurs when there is a full Moon at the same time that the Moon is at the closest point (*perigee*) in its orbit around Earth. Because it's closer than average, it looks bigger and brighter than at other times, even other full Moons. A supermoon may be only 7 percent larger than the Moon looks on average, so when you read in the newspaper that a coming supermoon will be 14 percent larger, what the article means is that the supermoon will be 14 percent larger than the Moon appears when it *looks as small as it ever does*. That happens when it's at its greatest distance from Earth (*apogee*). We used no such term when I learned and later taught astronomy, but lately "supermoon" has caught on in the media.

- A *blue moon* looks like other full Moons, and it's not blue in color. It's just a bit rare, as in the common expression "once in a blue moon." Ordinarily, there are 12 full Moons in a year (1 per month and 3 per 3-month season). But now and then, we have a year with 13 full Moons. That year has a month with two full Moons and a season (including that month) with four full Moons. No official definition of a blue Moon exists, but some folks say that the second full Moon in a month is a blue Moon, and some say that the third full Moon of a season with four of them is the blue Moon. I prefer the first definition, but I won't see red if you use the second one.

- The *Moon illusion* occurs when you watch the Moon rising above the horizon. It seems to most viewers that the Moon looks bigger then than it does a few hours later when the Moon is high in the sky. Despite hundreds of scientific studies and many theories of the Moon illusion, scientists don't agree on what makes the Moon look larger when it isn't. But we do know that when you see the Moon illusion, it's something in your brain, not a property of the Moon. Just be glad that you get a better look at the Moon without a telescope when you see the illusion. The same Moon illusion occurs when the Moon is setting at the horizon.

Earth has phases, too! To see them, however, you need to head into space and look back at Earth from a distance. When folks on Earth see a beautiful full Moon, an observer standing on the near side of the Moon would enjoy a "new Earth," and when earthlings experience a new Moon, viewers on the Moon would see a "full Earth."

In the shadows: Watching lunar eclipses

A lunar eclipse occurs when a full Moon is exactly on the line from the Sun to Earth. The Moon is then in Earth's shadow, or the *umbra*. You can safely look at a lunar eclipse, as long as you don't bump into something in the dark or stand in the road.

During a total eclipse of the Moon, you can still see the Moon, although it's immersed in Earth's shadow (see Figure 5-5). No direct sunlight falls on it, but some light from the Sun gets bent around the edges of Earth's atmosphere (as visible from the Moon) and falls on the Moon. The sunlight gets strongly filtered as it passes through our atmosphere, so mostly the red and orange light gets through. This *earthshine* effect differs from one lunar eclipse to the next, depending on meteorological conditions and the clouds in Earth's atmosphere. The totally eclipsed Moon, therefore, can look a dull orange, an even duller red, or a very dark red (sometimes called a Blood Moon). Sometimes you can barely make out the eclipsed Moon at all.

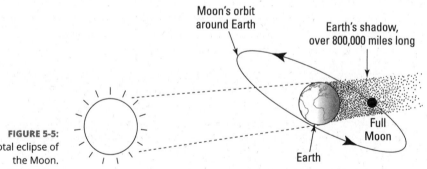

FIGURE 5-5:
A total eclipse of the Moon.

Dinah L. Moché/Astronomy: A Self-Teaching Guide, Seventh Edition

The dates for upcoming total lunar eclipses through the year 2028 are

January 31, 2018

July 27, 2018

January 21, 2019

May 26, 2021

May 16, 2022

November 8, 2022

March 14, 2025

September 7, 2025

March 3, 2026

December 31, 2028

TIP

To prepare for the upcoming eclipses, you can find plenty of information on the exact times and on the part of Earth where the eclipse will be visible. Take a look in *Astronomy* or *Sky & Telescope* magazines and on their websites (www.astronomy.com and www.skyandtelescope.com) as the eclipse dates approach.

Total eclipses of the Moon are as common as total eclipses of the Sun, but you see them more often at any given place because a total eclipse of the Sun is visible only along a narrow band on Earth called the *path of totality.* But when Earth's shadow falls on the Moon, you can see the eclipsed Moon from all over half of Earth, wherever night has fallen.

Partial eclipses aren't quite as interesting. During a *partial eclipse,* only some of the full Moon falls within Earth's shadow. The Moon just appears to be at a different phase. If you don't know that an eclipse is in progress — or that the full Moon phase is underway — you aren't aware that a unique astronomical event is occurring. You may simply pass it off as a quarter or crescent Moon. But if you keep looking for an hour or so, you can see the full Moon come out from Earth's shadow.

Cultivating an interest in the occult(ations)

As the Moon moves across the background stars while orbiting Earth, it sometimes eclipses a bright star. But we don't call it an eclipse. To astronomers, the event is a *lunar occultation.*

When a bright star is due to be occulted, you find notices of the event on the *Sky & Telescope* and *Astronomy* magazine websites and other sources of observing information I describe in Chapter 2. The first thing you need to know, which the notices provide, is where the occultation will be visible. If it's visible where you live, look for it at the specified time. Depending on the details, you may see the Moon occult the star and then move past so that the star reappears. Or you may be well placed to see only the star disappear behind the Moon, called *immersion,* or the star come out from behind the Moon, known as *emersion.* Depending on how bright the star is, you can watch the event by eye, or you may need binoculars or a small telescope. A star can be hard to see right next to the Moon because the Moon is so bright.

Besides enjoying the sight of an occultation, you can try to record the exact time of immersion and/or emersion from your location. These times depend on several factors:

>> The motion of the Moon

>> The position of the star

>> Your geographic location

>> The exact shape of the limb of the Moon behind which the star gets occulted and the shape of the opposite limb, where the star reemerges

If there's a high crater edge or lunar mountain at one of those spots, the star will wink out sooner or reappear a little later than calculated. (Head to the following section for info on craters and lunar mountains.)

The International Occultation Timing Association (IOTA) maintains the prime website for lunar occultation observers and also for astronomers who observe *asteroidal occultations,* meaning events when an asteroid passes in front of a star (which I describe in Chapter 7). Visit IOTA at `www.lunar-occultations.com/iota/iotandx.htm`

Hard rock: Surveying lunar geology

The entire Moon is pockmarked with craters of every size, from microscopic pits to basins hundreds of miles in diameter. The largest is the South Pole–Aitken Basin, which is about 1,600 miles (2,600 kilometers) across. Objects (asteroids, meteoroids, and comets) that struck the Moon — very long ago, for the most part — caused these craters. The microscopic craters, which scientists have found on rocks brought back by astronauts from the surface of the Moon, are caused by micrometeorites — tiny rock particles flying through space. All the craters and basins are known collectively as *impact craters,* to distinguish them from volcanic craters.

The Moon has experienced volcanism, but it took a form different from Earth's. The Moon has no *volcanoes,* or large volcanic mountains with craters at the top. But it does have small volcanic domes, or round-topped hills like ones that occur in some volcanic regions on Earth. In addition, sinuous channels on the lunar surface (called *rilles*) appear to be lava tubes, also a common landform in volcanic areas on Earth (such as Lava Beds National Monument in northern California). Most notably, the Moon has huge lava plains that fill the bottoms of the large impact basins. These lava plains are called *maria,* the Latin word for "seas." (When you look up and see the Man in the Moon, the dark areas that make up some of his features are the maria.)

Some early scientists thought the maria could be oceans. But if they were oceans, you'd be able to see bright reflections of the Sun from them, just as you do when you look down at the sea from an airplane during the day. The larger, bright areas in the Man in the Moon are the *lunar highlands,* which are heavily cratered areas. The maria have craters, too, but fewer craters per square mile than the highlands, which means that the maria are younger. Huge impacts created the basins where the maria are located. These impacts obliterated preexisting craters. Later the basins filled with lava from below, wiping out any new craters that had formed after the huge impacts. All the craters you can see in the maria now are from impacts that happened after the lava froze.

Moon ice? That's nice

A so-called lunar soil, consisting of fine rock dust, covers the surface of the Moon. It comes from countless impacts of meteoroids and asteroids that have struck the Moon for ages, making craters and pulverizing rock. Frozen molecules of water are stuck to the dust particles in many cases, especially in the bottoms of craters near the poles of the Moon. In the vicinity of those craters, the Sun is never high in the sky, and the crater bottoms are shadowed by the crater rims. These areas are the coldest places on the Moon. The temperature in at least one south pole crater gets below −400°F. But don't plan on scooping up Moon dust to make a refreshing glass of lunar ice water. Along with the water molecules, silver and mercury atoms are frozen to the dust, and you don't want them sliding down your throat.

Ready to observe the Moon's near side and find out the scoop on the Moon's far side? Check out the following sections.

Observing the near side

The Moon is one of the most rewarding objects to observe. You can see it when the sky is hazy or partly cloudy, and at times it's visible during the day. You can see craters with even the smallest telescopes. And with a high-quality small telescope, you can enjoy hundreds and maybe thousands of lunar features, including those mentioned earlier in the chapter and the following:

>> **Central peaks:** Mountains of rubble thrown up in the rebound of the lunar surface from the effects of a powerful impact. Central peaks are found in some, but not all, impact craters.

>> **Lunar mountains:** The rims of large craters or impact basins, which may have been partly destroyed by subsequent impacts. Parts of their walls remain standing alone like a range of mountains, although not the type of mountain you see on Earth.

>> **Rays:** Bright lines formed by powdery debris thrown out from some impacts. They extend radially outward from young, bright impact craters, such as Tycho and Copernicus (see Figure 5-6).

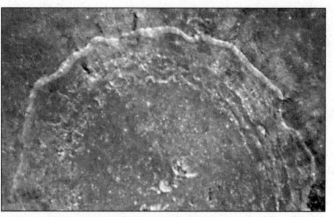

FIGURE 5-6:
A close-up view of the lunar crater Copernicus from the Hubble Space Telescope.

Courtesy of John Caldwell (York University, Ontario), Alex Storrs (STSci), and NASA

TIP

If you want to distinguish one crater, rille, or lunar mountain range from another as you look through your telescope, you need a Moon map or a set of lunar charts. These inexpensive items are available from astronomy and other scientific hobby supply houses and sometimes from map stores. Here are two good sources for these maps:

>> Orion Telescopes & Binoculars (www.telescope.com) sells maps and helpful guidebooks for lunar observing.

>> The website www.skyandtelescope.com offers a variety of Moon maps and globes in its shop. A beginner should consider the "Mirror-Image Moon Map (Laminated)" that costs about $6 (as of this writing).

Remember, these maps show only one side of the Moon: the lunar near side. You don't need a map of the far side of the Moon because you can't see it from Earth. But the far side is shown on globes for the benefit of armchair observers.

TIP

Start your exploration of the Moon by looking for a few of the most prominent maria and craters. You can easily spot many craters on the Moon with a small telescope and a Moon map. With a little more difficulty, you can identify some of them with a good pair of binoculars and in a few cases even by eye. Binoculars are excellent for scanning the lunar maria, which are readily visible with the eye as well. Start your Moon watching by looking for my choices for the top three maria:

Mare Crisium, the Sea of Crises; Mare Tranquilitatis, Sea of Tranquility (where Neil Armstrong and Buzz Aldrin landed in 1969); and Oceanus Procellarum, Ocean of Storms (the largest lunar mare). I list my top five craters in Table 5-1. You'll find these mares and craters marked on any good Moon map.

TABLE 5-1   **Top Lunar Craters for Your Viewing Pleasure**

Name	Diameter	Comments
Aristarchus	25 miles (40 kilometers)	Very bright and with bright rays
Copernicus	58 miles (93 kilometers)	Bright rays
Grimaldi	108 miles (174 kilometers)	Very dark crater floor
Plato	68 miles (109 kilometers)	Dark floor
Tycho	53 miles (85 kilometers)	Bright rays; in the lunar highlands

For almost anything you want to see on the Moon, the best viewing time is when the object is near the *terminator,* which is the dividing line between bright and dark. Details of lunar features are most evident when features are just to the bright side of the terminator.

During a month, which is approximately the period of time from one full Moon to the next, the terminator moves systematically across the lunar near side so that, at one time or another, everything you can see on the Moon is close to the terminator. Depending on the time of the month, the terminator is either the place on the Moon where the Sun rises or the place where the Sun sets. As you know from experience on Earth, shadows extend farther during sunrise or sunset and continually shrink as the Sun gets higher in the sky. The length of the shadow when the Sun is at a known altitude is related to the height of the lunar feature that casts it. The longer the shadow, the taller the feature.

Taking a shine to earthshine

When you observe the Moon, you may notice that the part on the dark side of the terminator isn't always pitch black. You may see a dim glow even though the Sun isn't shining there. The glow is earthshine, similar to the reddish glow on the lunar surface during a total eclipse of the Moon that I describe earlier in this chapter. *Earthshine* is sunlight that gets through Earth's atmosphere, where it's reddened (like the Sun at sunrise or sunset) and also bent slightly. It's bent enough to land on the Moon and cast a dim glow. You can see earthshine most easily when there's a crescent Moon, and you will never see it on a full Moon.

SKETCHING THE MOON

You can view or photograph the Moon and enjoy spotting maria, craters, and other topographic features marked on your lunar map. However, some amateur astronomers have a different hobby: They make drawings of lunar features seen through their telescopes. If you're artistically inclined, sketching lunar features may be an ideal astronomical hobby for you. The Moon is the only celestial object on which you can easily make out surface relief. You can do three-dimensional sketches of the lunar landscape, just as you may draw the view from your home. You can't do that for any other body unless you work from a space probe's close-up photo, not your own observations. *Sketching the Moon: An Astronomical Artist's Guide,* by Richard Handy, Deidre Kelleghan, Thomas McCague, Erika Rix, and Sally Russell (Springer), is a wonderful guide. It gives illustrated examples and step-by-step instructions for making your own telescopic drawings of the Moon by taking account of the interplay of light and shadow on the stark lunar surface.

WARNING

About the worst time to look at nearly anything on the Moon is during a full Moon. During a full Moon, the Sun is high in the sky on most of the lunar near side, so the shadows are few and short. The presence of shadows cast by features on the Moon helps you understand the *surface relief* — the way landforms extend above or below their surroundings.

Joining the dark side

You don't need a chart of the far side to help you observe the Moon because you can't see the far side. Our view is limited because the Moon is in *synchronous rotation,* meaning that it makes exactly one turn on its axis as it makes one orbit around Earth (the orbital period of the Moon is about 27 days, 7 hours, and 43 minutes). So the Moon always has the same hemisphere facing Earth.

Astronomy supply houses and science stores sell Moon globes that depict the features of the entire Moon, meaning the lunar near side and the far side. The Soviet space program first photographed the far side of the Moon, which it did by snapping pictures with a robotic spacecraft very early during the Space Age. Since then, many different spacecraft, including the Lunar Orbiters, Clementine, and the Lunar Reconnaissance Orbiter, have thoroughly mapped the Moon.

Being a lunar citizen scientist

You can study the Moon without a telescope and from the comfort of your home. Just surf to cosmoquest.org/x/science/moon/, click "Start Mapping," and take the tutorial. You'll learn to identify small features on the Moon in photographs from NASA's Lunar Reconnaissance Orbiter and perhaps contribute to lunar science.

BRING YOUR SUNSCREEN, OXYGEN SUPPLY, AND PARKA

When the Sun is up, the temperature on the lunar surface goes up to as much as 243°F (117°C), but at night, it drops to around –272°F (–169°C). These extreme temperature changes are due to the absence of any meaningful atmosphere to insulate the surface and reduce the amount of heat it loses at night. The Moon has no liquid water. The surface is too hot, too cold, and too dry to sustain life as we know it. And there's no air to breathe.

Quite an impact: Considering a theory about the Moon's origin

Scientists know a lot about the ages of the rock in different terrains and parts of the Moon. They acquired the data with radioactive dating of samples from the hundreds of pounds of lunar rocks that the six crews of NASA Apollo astronauts — who landed on the Moon at different times from 1969 through 1972 — brought back to Earth.

Before the Apollo Moon missions, several top experts confidently predicted that the Moon would be the Rosetta stone of the solar system. With no liquid water to erode the surface, no atmosphere worth mentioning, and no active volcanism on the Moon, they thought the surface should include plenty of primordial material from the birth of the Moon and the planets. But the Apollo lunar samples threw rocks on their theory.

When a rock melts, cools, and crystallizes, all its radioactive clocks are reset. Radioactive isotopes begin producing fresh daughter isotopes that become trapped in the newly formed mineral crystals. The Apollo Moon rocks show that the whole Moon — or, at least, its crust down to a considerable depth — was melted well after 4.6 billion years ago. The very oldest surface rocks on the Moon are *only* 4.5 billion years old. The difference between 4.6 and 4.5 billion years is 100 million years. And unlike the minerals in Earth rocks, which contain water bound up into the mineral structures, the Moon rocks are almost bone dry.

The origin theory that has emerged to explain all this evidence, and to avoid the objections scientists posed against previous theories, is the *Giant Impact theory*. According to this theory, the Moon consists of material blasted out of the mantle of Earth by a huge object — with up to three times the mass of Mars — that struck young Earth a glancing blow. Some of the rock from the mantle of that long-vanished impacting object also was incorporated into the Moon, according to the theory, and Earth ended up with its axial tilt.

The giant impact on young Earth knocked all this material up into space as a vapor of hot rock. It condensed and solidified like snowflakes. The snowflakes knocked into each other and stuck together, and before you knew it, the Moon had formed. It came together in powerful impacts of the last big pieces of accumulated rock, with the heat from each impact melting the rock.

All the impacts that caused the craters that we now see on the Moon happened later, and most of them date back to more than 3 billion years ago.

The Moon is less dense than Earth as a whole and about as dense as Earth's mantle (the layer beneath the crust and above the core), according to this theory, because it was made from mantle material. (Density is a measure of the amount of mass that's packed into a given volume. If you have two cannonballs of the same size and shape, they have the same volume. But if one ball is made of lead and one is made of wood, the lead ball is heavier and has a higher density.) This theory predicts that the Moon shouldn't have much of an iron core, if any. And a small core in a small object (meaning the Moon) should have cooled and frozen long ago if it ever contained liquid iron. Lunar researchers suspect that, nevertheless, the Moon has an iron core and that it may be partly molten.

The Giant Impact theory is currently our best guess. Unfortunately, we have no test for it at this time. For example, the theory predicts no special kind of rock that we can look for in the hundreds of pounds of lunar rocks that the Apollo astronauts collected. Some astronomers suspect that rocks in the biggest crater on the Moon were knocked out from a great depth by the colliding asteroid that formed the crater. Those rocks, in the South Pole–Aitken Basin, may come from a depth so great that rock there didn't melt when the Basin formed; the rocks might be samples of the lunar mantle, the layer beneath the crust. Studies of the rocks could tell scientists whether the Giant Impact theory is accurate. The South Pole–Aitken Basin is the largest crater on the Moon — or anywhere else in the solar system — and is also the topic of big disagreements. Some experts think the rocks there don't come from deep enough to test the theory. Worse yet, plans to explore the Basin with astronauts or robotic craft are on hold as NASA struggles to devise a new Moon program.

If scientists verify the Giant Impact theory, it will have a giant impact on science, but don't hold your breath — this could take many moons. In fact, a few astronomers recently suggested that Earth did have many moons long ago and that they merged into the one big Moon that we enjoy today.

- » **Checking out Venus, hot and stuffy with acid rain**

- » **Discovering Mars, the planet we search for water and life**

- » **Understanding what sets Earth apart**

- » **Finding and observing our neighboring planets**

Chapter **6**

Earth's Near Neighbors: Mercury, Venus, and Mars

You can spot Earth's neighboring terrestrial (or rocky) planets Mercury, Venus, and Mars with the naked eye and inspect them with your telescope. But they tantalize you by revealing only a little of their nature in that way. Most of what scientists know about their physical properties, geologic forms, and histories is based on images and measurement data that interplanetary spacecraft sent back to Earth.

Two NASA spacecraft have visited Mercury, one flying past it three times and one orbiting the planet. Several probes have visited, orbited, and even landed on Venus. Mars has been the target of numerous probes, landers, and robot rovers. Some have orbited Mars, some have landed on it, and some, unfortunately, have crashed or even missed the planet. Getting there safely isn't brain surgery, but it is rocket science, and it's hard.

In this chapter, I give you fascinating details about (and handy tips for viewing) Earth's closest neighbors in the solar system.

Mercury: Weird, Hot, and Mostly Metal

Astronomers have an ironclad case that Mercury is weird. It's not mostly rock, like Earth (or like the Moon, Mars, and Venus), but it's mostly metal — a big ball of iron with a thin skin of rock. On Earth, the iron core (see Chapter 5) extends just over halfway from the center to the surface. But on Mercury, the metal core extends 85 percent of the way to the surface, and it's at least partially molten. Mercury also has one layer that's unlike the layers inside any other planet: a zone of solid iron and sulfur between the iron core and the rocky surface.

Mercury's weirdness is plain to see on its surface as well. A great many impact craters pock the surface, as on the Moon (see Chapter 5), but many of Mercury's craters are tilted, as though the ground shifted after they formed. The largest crater on Mercury, the Caloris Basin (960 miles [1,545 kilometers] across) is strange, too: Much of the crater *bottom* is raised above the rim. Try to figure out how that happened!

TIP

Much of what we know about Mercury comes from NASA's MESSENGER probe, which spent more than 10 years in space and orbited Mercury from 2011 to 2015. Those studies built on the investigations by an earlier spacecraft, Mariner 10, which made three flybys of the planet in 1974 and 1975. You can see some of the best pictures of Mercury in the Highlights Collection page of the MESSENGER website at the Johns Hopkins University at messenger.jhuapl.edu/Explore/Images.html#highlights-collection.

TECHNICAL STUFF

Why is MESSENGER all in caps? NASA loves acronyms, meaning words or names of things that are formed from initials. In fact NASA, itself is an acronym for National Aeronautics and Space Administration. MESSENGER stands for MErcury Surface, Space ENvironment, GEochemistry, and Ranging. Well not all the letters in that name are initials, but they're close enough for government work.

Here are some of the key facts about Mercury:

>> Mercury has long, winding ridges that cut across impact craters and other geologic features. The ridges were probably caused by shrinkage of the crust as Mercury contracted while cooling from a molten state. The diameter of the planet may have shrunk by as much as 8 miles (13 kilometers).

>> Mercury has fewer small craters than the Moon, in proportion to the number of large craters.

>> Highly cratered highlands are present on Mercury, as on the Moon (Mercury has no known moon of its own). But unlike on the Moon, Mercury's highlands are interrupted by gently rolling plains.

>> On the antipode of the Caloris Basin, exactly opposite Caloris on the other side of Mercury, is a region of broken terrain. The collision that caused Caloris must have generated powerful seismic waves, which traveled through Mercury and around its surface, converging at the antipode with a catastrophic effect.

>> Mercury has a global magnetic field, generated by a natural dynamo in its molten iron core, much like Earth, except that Earth's magnetic field is 100 times stronger. (Mars, Venus, and the Moon all lack a global magnetic field.)

>> Mercury has huge swings in temperature, from as much as 870°F (466°C) during the day to as low as –300°F (–184°C) at night.

>> Ice is present in the bottoms of craters near Mercury's North Pole, which are always in shadow and very cold.

>> Lava streamed across Mercury long ago, carving out valleys.

The chemistry of Mercury is strange, with lots of the volatile (easily evaporating) elements potassium, sodium, and sulfur on the surface. That finding challenges astronomers who try to explain how Mercury got to be mostly iron. Previous theories postulated that Mercury once had a larger proportion of rock than it does now. If so, it would have been more like Earth and the Moon. The theories proposed various ways of blasting away much of the original outer rock of Mercury, but any force so powerful would also have vaporized the potassium, sodium, and sulfur that now pervade the Mercury surface.

The more astronomers have learned about Mercury, the more it puzzles us. Check out the color section of this book to see an image of Mercury.

Dry, Acidic, and Hilly: Steering Clear of Venus

Venus never sees a clear day; the planet is perpetually covered from equator to pole by a 9-mile-thick (15-kilometer) layer of clouds made up of concentrated sulfuric acid. And the surface has no relief from the heat: Venus is the hottest planet in the solar system, with a surface temperature of 870°F (465.5°C) that stays about the same from equator to pole, day and night.

And if the heat seems bad, check out the barometric pressure: It measures about 93 times the pressure at sea level on Earth. But forget about seas; you won't find any water on Venus. You can complain about the heat, but not the humidity — it's a dry heat, like in Arizona.

The bad news about the weather on Venus is that a perpetual rain of sulfuric acid falls all over the planet. The good news is that this rain is a *virga*, meaning rain that evaporates before it hits the ground. So if you're ever grounded on Venus, at least you won't have an acid bath.

The high temperature on Venus results from an extreme case of the *greenhouse effect*. In simple terms, the thick atmosphere (which is more than 95 percent carbon dioxide, with no oxygen) and clouds let through most of the sunlight that falls on Venus, which heats the surface and the air near the ground. The warm surface and air release heat in the form of infrared rays. On Earth, a lot of such infrared radiation escapes into space, cooling the ground at night. But on Venus, the carbon dioxide atmosphere traps the infrared, so the planet gets extremely hot.

Many of the excellent images of the surface of Venus that you can find on NASA websites (and others) aren't photographs at all. What you see are detailed radar maps, notably from NASA's Magellan spacecraft. The clouds on the planet block the view of telescopes on Earth and of any camera on a Venus-orbiting satellite. The cloud tops are at an altitude of 40 miles (65 kilometers), much lower than where a satellite can operate.

TIP

The few images we have from Venus lander probes, launched by the former Soviet Union, show areas of flat rock plates separated by small amounts of soil. The plates resemble areas of hardened basalt lava flows on Earth. But on Venus, the surface appears orange because the thick cloud cover filters the sunlight. You can find some of the old Soviet pictures of rocky landscapes at the National Space Science Data website, nssdc.gsfc.nasa.gov/photo_gallery, where you click on "Venus" and then on "Surface Views." (You can see an image of Venus in the color section of this book, too.)

Flat plains that are volcanic lowlands with *rilles* (the winding canyons left by lava flows) cover the vast majority of Venus (about 85 percent). This territory includes the longest-known rille in the solar system, Baltis Vallis, which stretches across Venus for about 4,230 miles (6,800 kilometers). Cratered highlands and deformed plateaus are also present.

Not as many craters dot Venus as you may expect, based on the number you see on Earth's Moon (Venus has no known moon) and on Mercury. No small craters exist, and there aren't many large craters because Venus's surface was flooded with lava or reworked by volcanism (the eruption of molten rock from within a planet) after its bombardment by impacting objects had mostly ended. This flooding or reworking erased all or most of the early craters. Few large objects have struck Venus since the early craters were destroyed, and small objects don't make many craters on Venus: Aerodynamic forces in the thick Venus atmosphere impede and destroy objects capable of making craters up to 2 miles (3 kilometers) in diameter.

Huge volcanoes (one or more of which may be active now, or at least may have spewed lava within past thousands of years, not just millions or billions of years ago) and mountain ranges cover the surface of Venus. But nothing there resembles the nonvolcanic mountains on Earth (like the Rocky Mountains in the western United States or the Himalayas in Asia), which are caused by one crustal plate pushing into another. And Venus has no chains of volcanoes (like the Pacific "Ring of Fire"), which rise at the edges of plates. Plate tectonics and continental drift don't occur on Venus as they do on Earth.

The European Space Agency probe Venus Express arrived at Venus in April 2006 and orbited it until May 2014. It studied the hot planet's atmosphere in great detail, gathered tantalizing evidence of recent volcanism, and found that the rotation of Venus may be slowing down. You can find images from Venus Express at www.esa.int/spaceinimages/Missions/Venus_Express/. Venus pictures from NASA spacecraft including Magellan, Galileo, and the Hubble Space Telescope are in the Planetary Photojournal site at photojournal.jpl.nasa.gov (just click on "Venus").

Red, Cold, and Barren: Uncovering the Mysteries of Mars

Scientists have mapped the surface of Mars and measured the altitudes of mountains, canyons, and other features with a high degree of accuracy. You can find the National Geographic chart of the entire planet at the NASA website (tharsis.gsfc.nasa.gov/ngs.html). The map is based on data from two instruments carried on the Mars Global Surveyor (MGS), a satellite that operated in orbit around Mars from 1997 to 2006. One instrument, a laser altimeter, bounced pulses of light off the Mars surface to measure the altitudes of the surface features that reflected the light. The other instrument, a camera, photographed landforms.

While MGS was still operating, another NASA probe, the Mars Odyssey, arrived and began orbiting the planet in October 2001. As this book goes to press, Odyssey is still going strong. It's now the longest operating spacecraft at Mars. It has found caves and evidence of widespread ice and also salt deposits on Mars. You can't lick that.

TIP

The European Space Agency doesn't get as much publicity as NASA, so you may not know that the Europeans have a Mars Express satellite that began orbiting the red planet on December 25, 2003. You can see splendid images of layered structure on the North Polar Cap of Mars and other scenes from this spacecraft and enjoy video flybys of Mars's terrain created from the spacecraft images at www.esa.int/ Our_Activities/Space_Science/Mars_Express. There's even music that accompanies the videos, but they're more rocks than rock and roll.

Even though scientists have accurately mapped Mars with satellites and have explored parts of the Martian surface with lander probes and robot rovers, the planet holds many mysteries that they want to solve. In the following sections, I cover theories about water and life on Mars. (For even more on Mars, be sure to check out an image in the color section of this book.)

Where has all the water gone?

The topographic map of Mars shows that most of the Northern Hemisphere is much lower than the Southern Hemisphere. Some astronomers believe that this arid northern region, bounded by features that resemble an ancient shoreline, was an ocean long ago. Scientists used ground-penetrating radar mounted on Mars Express to peek under the possible shoreline region. In 2012, they reported that the subsurface layer resembles *sediment* (meaning broken rock, soil, and/or sand that settles to the bottom of a body of water). The sediment (if that's what it is) presumably was deposited in the suspected ancient sea. Some experts are still not convinced that Mars had an ocean, but I'm a believer.

The northern lowland may or may not be the site of an ancient ocean, but even if it isn't, other evidence suggests that liquid water was once common on Mars:

>> Mars is cold and dry now, but there's a great deal of ice at the poles. By one estimate, enough ice is present to flood the entire planet to a depth of 100 feet (30 meters) if it melted. (The polar ice won't melt; Mars is just too cold.)

>> Some canyons on Mars look like a great flood carved them out long ago.

>> The Mars Reconnaissance Orbiter photographed many narrow linear features that run down steep slopes. They darken during the Martian summer and fade as the weather cools. They may be caused by salty water in the surface layer that melts and flows only when Mars is warm enough. (Saltwater stays liquid at lower temperatures than freshwater, so saltwater may flow on Mars when freshwater is solid ice.)

>> The soil in many parts of Mars contains substances that form under the influence of water, including minerals that occur in clays.

>> The planet has features that look just like dry riverbeds, with streamlined islands and pebbles that seemingly were rounded in a torrent. Such pebbles were imaged by the Mars Pathfinder lander probe and its little robot, Sojourner.

>> Mars Odyssey found indications of much water, presumably frozen, just beneath the surface over large areas of Mars.

>> NASA's Curiosity, the largest rover to traipse around Mars, identified the remains of what was once a freshwater lake.

The Mars atmosphere is mostly carbon dioxide (as is Venus's atmosphere), but the Martian atmosphere is much thinner than the atmospheres of Earth or Venus. Mars also has clouds of water-ice crystals, which resemble the cirrus clouds on Earth. In winter, some carbon dioxide from the Martian atmosphere freezes on the surface, leaving thin deposits of dry ice. The South Polar Cap is always covered in dry ice (water ice may lie below it). But water ice is seen at the North Pole when it's summer there and the dry ice has evaporated for the season. If Mars had an ocean in the past, the planet had to have been much warmer than it is today. If the carbon dioxide atmosphere was much thicker then, it would have trapped heat, as in the greenhouse effect on Venus that I describe earlier in this chapter. If Mars once had a warm atmosphere and an ocean, then carbon dioxide from the atmosphere would have dissolved in the water. Chemical reactions then would have produced carbonates (minerals composed of carbon and oxygen). This theory predicts that there are carbonate rocks on Mars. NASA's Mars Exploration Rover, Spirit, discovered carbonate rocks in 2010!

Mars experts have a variety of opinions, but I think the case is closed: Mars once had a warm climate and a whole lot of liquid water.

Currently, it gets comfortably warm in the daytime on Mars at the equator, where noontime temperatures can reach a balmy 62°F (16.6°C). However, don't stay the night — it can get down to −208°F (−133.3°C) after sunset. The seasons on Mars differ from Earth's seasons, too. As I explain in Chapter 5, Earth's seasons are caused by the tilt of Earth's axis with respect to the plane of Earth's orbit around the Sun, *not* by changes in Earth's distance from the Sun (which are negligible). On Mars, both the tilt of the planet's axis and the significant changes in its distance from the Sun from one place in Mars's orbit to another (because the orbit of Mars is more elliptical than Earth's almost-circular orbit) combine to produce "unearthly" seasons. Summer in the Southern Hemisphere on Mars is shorter and hotter than summer in the Northern Hemisphere, and winter in the Northern Hemisphere on Mars is shorter and warmer than winter in the Southern Hemisphere.

A magnetometer on MGS discovered long parallel stripes of oppositely directed magnetic fields frozen in the rocky crust of Mars. Mars doesn't have a global magnetic field today, but this finding may mean that it once had a global field that periodically reversed, just as Earth's field does (see Chapter 5). It may also mean that Mars endured a crustal process resembling the seafloor spreading on Earth and producing a similar pattern. The molten iron core on Mars should have frozen solid long ago, and indeed a new magnetic field is no longer generated, but nevertheless evidence indicates that the core may still be at least partially molten. The heat flow from the inside to the surface is so low that there's probably no volcanism still underway.

The volcanism that did occur on Mars produced immense mountains, such as Olympus Mons, the largest volcano in the solar system. It's about 370 miles (600 kilometers) wide and 15 miles (24 kilometers) high, or five times wider and almost three times higher than the biggest volcano on Earth, Mauna Loa. Mars also has many canyons, including the immense Valles Marineris (Mariner Valley), which is 2,490 miles (4,000 kilometers) long. Impact craters dot the surface, too. The craters are more worn down than those on Earth's Moon because much more erosion has occurred, possibly caused by water that supposedly produced great floods on Mars.

Does Mars support life?

People have many mistaken ideas about Mars, but some of the theories may actually be right; they just haven't been proven. These ideas all revolve around the possibility of life on Mars. Most of them are as improbable as the story about the future astronaut who returns from the planet: "Well, is there life on Mars?" the reporters demand. "Not much during the week," he says, "but on Saturday night. . . ."

Claims about life strike out

The discovery of the "canals" on Mars spawned the first widespread speculation about the possibility of life. Some of the most famous astronomers of the late 19th and early 20th centuries were among those who reported the canals. Planetary photography wasn't very useful in those days because the exposures were fairly long and atmospheric seeing (which I define in Chapter 3) blurred the images. So scientists believed that drawings by expert professional telescopic observers were the most accurate images of Mars. Some of these charts showed patterns of lines stretching and crisscrossing around the surface of Mars. Percival Lowell, an American astronomer, theorized that the straight lines were canals, engineered by an ancient civilization to conserve and transport water as Mars dried up. He concluded that the places where the lines crossed were oases.

Over the years, the idea of the "canals" and other reported indications of past or present life on Mars have struck out:

>> When the American spacecraft Mariner 4 reached Mars in 1965, its photographs showed no canals, a conclusion verified in much greater detail by images from subsequent Mars probes. Strike one.

>> Two later probes, the Viking Landers, conducted robotic chemical experiments on Mars to look for evidence of biologic processes such as photosynthesis or respiration. At first they appeared to have found evidence of biologic activity when water was added to a soil sample. But most scientists who reviewed the matter concluded that the water was reacting chemically with the soil in a natural process that doesn't involve the presence of life. More recent probes that roved on the surface, including Curiosity, likewise found no compelling evidence of life. Strike two.

>> Viking Orbiters also sent back images of the Mars surface as they revolved around the planet. The images show, at one location, a crustal formation that — to some folks — looks like a face. Although many natural mountain peaks and stone formations on Earth resemble the profiles of the famous rulers, Native American tribal chiefs, and others for whom they're named, some true believers claim that the "face on Mars" is a monument of some type, erected by an advanced civilization. Later, sharper images from MGS showed that this landform doesn't look like a face at all. Strike three for the advocates of life on Mars.

But the idea of life wasn't "out," despite the three strikes. In 2003, astronomers began finding traces of methane (sometimes called "swamp gas") on Mars. Methane breaks down rapidly into other substances under Mars conditions, so some astronomers suggested that fresh methane is being generated by a primitive form of life on Mars. (On Earth, for example, microbes called methanogens give off methane.) However, geological processes on Mars also may make methane, so experts are still sniffing around this mystery.

The search for fossil evidence

In 1996, scientists analyzed samples of a meteorite that they believed came from Mars after being knocked off the planet by the impact of a small asteroid or comet. The scientists found chemical compounds and tiny mineral structures that they interpreted as chemical byproducts and possible fossil remains of ancient microscopic life. Their work is controversial, and many subsequent studies contradict these conclusions. Based on current research, scientists can't make a persuasive case that supports the theory of past life on Mars, nor can they disprove it.

The only action to take is to search systematically on Mars for evidence of life, past or present, in the regions that make the most sense — places where large quantities of water appear to have been present in the past and where layers of sediment were deposited in ancient lakes or seas. These types of places hold the most fossils on Earth.

Differentiating Earth through Comparative Planetology

Mercury is a small world of extreme temperatures, with a global magnetic field like Earth's, but much weaker. Neither Venus nor Mars has such a magnetic field, although the two planets are similar to Earth in many other ways. But liquid water and life occur today only on Earth, as far as we know. What makes Earth different?

Venus, unlike Earth, has a hellish temperature. Venus is farther from the Sun than Mercury but is even hotter. The high temperature is due to an extreme *greenhouse effect*, the process by which atmospheric gases raise the temperature by absorbing outward flowing heat. Earth's atmosphere may once have contained large amounts of carbon dioxide, the way Venus's atmosphere does now. But on Earth, the oceans absorbed much of the carbon dioxide, so that gas couldn't trap as much heat in the atmosphere as it does on Venus.

Mars, on the other hand, is too cold to support life (as far as we can tell). Mars has lost most of its original atmosphere, and its current atmosphere isn't thick enough to produce a greenhouse effect sufficient to warm much of the surface above the freezing point of water often or for very long.

The three large terrestrial planets are like the bowls of cereal in the child's story of Goldilocks. Venus is too hot, Mars is too cold, but Earth is *just right* to support liquid water and life as we know it. (Little Mercury is also too hot.) Putting together the information on the basic properties of the terrestrial planets and their respective differences, scientists can conclude that

» Mercury is like the Moon on the outside (with lots of craters) but like Earth on the inside, with a molten iron core that generates a magnetic field.

» Venus is Earth's "evil twin." It's about the same size as Earth, but with deadly heat and pressure, an unbreathable atmosphere, and highly acid rain.

» Mars is the little Earth that cooled and dried up.

Earth is the Goldilocks planet — just right!

When you contrast the properties of the planets like this, you can draw conclusions about their respective histories and why those different histories have brought the planets to their present states. Think that way, and you're practicing what astronomers call *comparative planetology.*

Observing the Terrestrial Planets with Ease

You can spot Mercury, Venus, and Mars in the night sky with the aid of monthly viewing tips from astronomy magazines and their websites, a smartphone app, or a desktop planetarium program (see Chapter 2 for information on all these resources). Venus is especially easy to find because it's the brightest celestial object in the night sky other than the Moon.

TIP

Mercury is the planet that orbits closest to the Sun, and Venus is next. They both orbit inside the orbit of Earth, so Mercury and Venus are always in the same region of the sky as the Sun, as seen from Earth. Therefore, you can find these planets in the western sky after sunset or in the eastern sky before dawn. At those times, the Sun isn't far below the horizon, so you can see objects near and to the west of the Sun in the morning before sunrise, and you can see objects near but east of the Sun in the evening after sunset. Your motto as a Mercury or Venus spotter is "Look east, young woman" or "Look west, young man," depending on whether you skywatch at dawn or dusk and whether you're a fan of old Western movies.

A bright planet appearing in the east before dawn is commonly termed a *morning star,* and a bright planet in the west after sunset is an *evening star.* Astronomers know that these aren't stars, but that's what people call them. As Mercury and Venus move swiftly around the Sun, this week's morning star may be the same object as next month's evening star (see Figure 6-1).

In the following sections, I explain the best time to observe the terrestrial planets based on elongation, opposition, and conjunction — three terms that describe the planets' positions in relation to the Sun and Earth — and how to use this understanding in your observations of the terrestrial planets. (I list the planets in the order of ease of observation, starting with Venus, the easiest.)

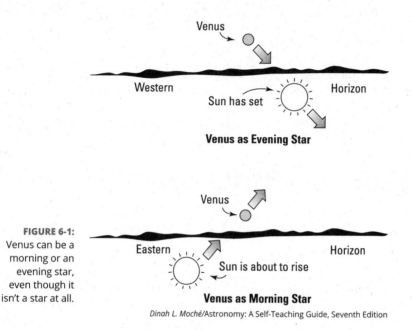

FIGURE 6-1:
Venus can be a morning or an evening star, even though it isn't a star at all.

Venus as Evening Star

Venus as Morning Star

Dinah L. Moché/Astronomy: A Self-Teaching Guide, Seventh Edition

Understanding elongation, opposition, and conjunction

Elongation, opposition, and *conjunction* are terms that describe a planet's position in relation to the Sun and Earth. You encounter these terms when you check listings of the planets' positions to plan your observations. Here's what the terms mean:

» *Elongation* is the angular separation between a planet and the Sun, as visible from Earth. Mercury's orbit is so small that the planet never gets more than 28° from the Sun. During some periods, it doesn't get farther than 18° from the Sun, making it hard to spot. Venus can get up to 47° from the Sun.

   Greatest western (or eastern) elongation occurs when a planet is as far from the Sun as it can get during a given *apparition* (a period of time when the planet is visible from Earth on successive nights). Some greatest elongations are greater than others because sometimes Earth is closer to the planet than at other occasions. Elongation is especially important when you observe Mercury because the planet is usually so close to the Sun that the sky at its position isn't very dark.

» *Opposition* occurs when a planet is on the opposite side of Earth from the Sun. Opposition never happens for Mercury or Venus, but Mars is at opposition about once every 26 months. This is the best time to observe the planet because it appears largest in a telescope. And at opposition, Mars is highest in the sky at midnight, so you can view it all night.

>> *Conjunction* occurs when two solar system objects are near each other in the sky, such as when the Moon passes near Venus as we see them. In reality, Venus is far beyond the Moon, but we see a conjunction of the Moon and Venus.

Conjunction has a technical meaning as well. Instead of describing positions in *right ascension* (the position of a star measured in the east–west direction) and *declination* (the position of a star measured in the north–south direction), astronomers sometimes use ecliptic latitude and longitude. The *ecliptic* is a circle in the sky that represents the path of the Sun through the constellations. *Ecliptic latitude* and *longitude* measure degrees north and south (latitude) or east and west (longitude) with respect to the ecliptic. (Don't worry; you don't need to use the ecliptic system when you observe the terrestrial planets. But knowing about it helps you understand the definitions of *superior* and *inferior conjunction* that follow.)

You need to master some tricky terminology to understand conjunctions and oppositions — namely, the labeling of planets as superior and inferior and the labeling of conjunctions as superior and inferior. A *superior planet* orbits outside the orbit of Earth (so Mars is a superior planet, for example). An *inferior planet* orbits inside the orbit of Earth (so Mercury and Venus are inferior planets — in fact, they're the only inferior planets).

When a superior planet is at the same longitude as the Sun as seen from Earth, it's on the far side of the Sun and is said to be at *conjunction* (see Figure 6-2). When that same planet is on the opposite side of Earth from the Sun (also shown in Figure 6-2), it's at *opposition*.

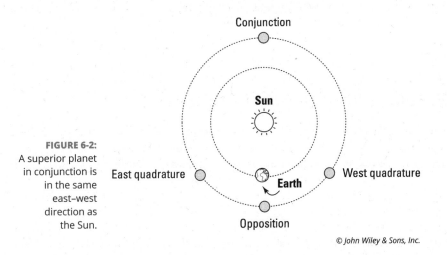

FIGURE 6-2: A superior planet in conjunction is in the same east–west direction as the Sun.

© *John Wiley & Sons, Inc.*

Conjunction is a bad time to observe a superior planet because it's on the far side of the Sun and also in nearly the same direction. So don't try to observe Mars at conjunction; you won't see it. The best time to observe Mars is at opposition.

A superior planet has conjunctions and oppositions, but an inferior planet has two kinds of conjunction and never has an opposition (see the diagram in Figure 6-3). When the inferior planet is at the same longitude as the Sun and is between the Sun and Earth, it's at inferior conjunction. And when the inferior planet is at the same longitude as the Sun but is beyond the Sun as seen from Earth, the planet is at superior conjunction.

If you can explain all that to your friends, you'll truly feel superior! Feel free to do the explaining "in conjunction" with Figures 6-2 and 6-3.

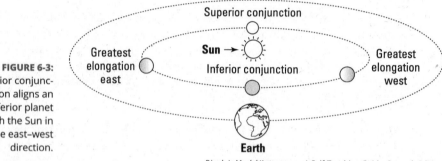

FIGURE 6-3: Inferior conjunction aligns an inferior planet with the Sun in the east–west direction.

Dinah L. Moché/Astronomy: A Self-Teaching Guide, Seventh Edition

TIP

You can see Venus best when it's between greatest elongation and inferior conjunction, when it looks brightest, but Mercury is too close to the Sun as seen from Earth for viewing at inferior or superior conjunction. The best time to see Mercury is when it's at greatest elongation.

Viewing Venus and its phases

The simplest planet to find is Venus. The second rock from the Sun is so bright that people with no knowledge of astronomy frequently notice it and call radio stations, newspapers, and planetariums to ask what "that bright star" is.

When scattered clouds move from west to east in front of Venus, inexperienced viewers often misunderstand the scene. Folks think that Venus (which they don't recognize) is moving rapidly in the opposite direction of the clouds. Due to its brightness and the mistaken impression that it's moving rapidly behind a cloud deck, people often report Venus as an unidentified flying object. It isn't. Astronomers know it well.

When you become familiar with Venus, you may be able to spot it in broad daylight. Often Venus is bright enough that, if the sky is clear, you can spot it in the daytime by using *averted vision*. In other words, you can glimpse it "out of the corner of your eye." For some reason, you may be able to spot a celestial object easier with averted vision than if you look right at it. The ability to spot things with averted vision may be a survival trait; it makes it a little harder for an enemy or a predator to sneak up on you from the side.

A small telescope can show you Venus's most recognizable features: its phases and changes in apparent size. Venus has phases similar to Earth's Moon (see Chapter 5), and for the same reason: Sometimes part of the Venus hemisphere that faces the Sun (and is, therefore, bright) is directed away from Earth, so a telescopic view of Venus shows a partly illuminated and partly dark disk.

The dividing line between the bright and dark parts of Venus is called the *terminator*, just as on the Moon. Don't worry: This terminator is just a perfectly safe imaginary line on Venus (see Chapter 5).

WAIT JUST AN ARC MINUTE (OR SECOND)

Scientists measure apparent sizes in the sky in angular units. Something that goes all the way around the sky, such as the celestial equator, is 360° long. The Sun and the Moon, in comparison, are each about a half-degree across. The planets are much smaller, so smaller units are necessary to describe them. A degree is divided into 60 minutes of arc, and a minute of arc — called an *arc minute* or *arc min* — is divided into 60 seconds of arc — usually called an *arc second* or *arc sec*. A degree is made up of 3,600 (60×60) seconds of arc. In many astronomy books and articles, a single prime symbol (') represents a minute of arc, and a double prime (") represents a second of arc. Readers often mistake these symbols as abbreviations for feet and inches. You can tell when a clueless copy editor has had the last cut at an astronomy article because you see a statement like "The Moon is about 30 feet in diameter."

Venus is actually only about 5 percent smaller in diameter than Earth. Its apparent size, or *angular diameter*, ranges from about 10 arc sec when Venus is farthest away (and with a full Moon shape) to about 58 arc sec in diameter when it's closest (and is a narrow crescent).

Here's an interesting factoid for you: An arc second equals the apparent size of a dime (U.S. ten-cent coin) seen from a distance of about 2 miles (3 kilometers).

As Venus and Earth orbit the Sun, the distance between the two planets shifts substantially. At its closest to Earth, Venus is a mere 25 million miles (40 million kilometers) away; at its farthest, it's a full 160 million miles (257 million kilometers) distant. What's important here is the proportional change: At closest approach, Venus is about six times nearer to Earth than at its most distant position. Therefore, it looks six times bigger through a telescope.

What you don't see when you view Venus are striking features, like the Man in the Moon. Venus is totally covered by thick clouds, and all you can see is the top of the clouds. Venus is so bright because it orbits relatively close to both the Sun and Earth and because it has a nice, bright reflecting cloud layer. But sometimes you may be able to discern the horns of the Venus crescent extending farther into the dark side than predicted for the phase on that day. In this case, you see some sunlight that has bounced around in Venus's atmosphere and passed beyond the terminator into the side of the planet where night has fallen.

Images of Venus with striking cloud patterns, like those you see in books, were made in ultraviolet light where the patterns show up. Ultraviolet light doesn't pass through our atmosphere (hurray for the ozone layer, which blocks this hazardous radiation), so you can't view Venus in it. In fact, you can't see ultraviolet light anyway; it's invisible to the human eye. But telescopes on satellites and space probes above or beyond the atmosphere can take ultraviolet pictures. Some Venus images on websites that I recommend earlier in the chapter were made in infrared light or radio waves; our eyes can't see those wavelengths either.

On rare occasions, observers report a pale glow on the dark part of Venus. This glow, called the *ashen light,* is sometimes real and sometimes a trick of the imagination. After centuries of study, experts still can't explain the ashen light, so some of them even deny that it exists. But with luck, you may see it. People claim to see other features on Venus through their telescopes, but almost all the reports are wrong. Experiments show that the reports are usually due to a psychological effect: If people view a featureless white globe from a distance, they may discern patterns that don't exist.

Watching Mars as it loops around

Mars is a bright red object, but it isn't nearly as dazzling as Venus. So check your sky maps to make sure you don't mistake a bright red star, such as Antares in Scorpius (whose name means "rival of Mars"), for the red planet.

The great advantage in observing Mars is that, when it appears in the night sky, it often remains visible for much of the night, unlike Mercury and Venus (which set fairly soon after sunset or rise only shortly before dawn). You usually have time for dinner and the nightly news before you head into the backyard to check on Mars.

TRY TO CATCH A PLANETARY TRANSIT

On rare occasions, you can see a *planetary transit,* when a planet passes right in front of the Sun and looks like a tiny black disk against the bright solar surface. Only the inferior planets (Mercury and Venus) transit the Sun because only they come between Earth and the Sun.

You can observe a transit of Mercury with a small telescope, but you must use the procedures for safe solar viewing that I explain in Chapter 10. Otherwise, you risk incurring severe eye damage and even blindness. It may be a good idea to check out whether your local planetarium, museum, or astronomy club is holding a transit-viewing event (they often do). Then you can look with equipment set up by experts. The next two transits of Mercury are on November 11, 2019, and November 13, 2032.

Transits of Venus are more impressive than those of Mercury because Venus has a bigger disk, but Venus transits are also much rarer. Unfortunately, if you haven't already seen a Venus transit, I think you never will. The last one was in June 2012 and the next transit of Venus won't occur until December 2117. Leave a note for your descendants to be sure and watch.

TIP

With a small telescope, you can spot at least a few dark markings on Mars. The best periods to see the features last a few months but occur only about every 26 months, when Mars is at opposition. At opposition, Mars looks biggest and brightest, and you can see details more easily on its surface.

The upcoming oppositions of Mars are in

> *July 2018*
>
> *October 2020*
>
> *December 2022*
>
> *January 2025*
>
> *February 2027*
>
> *March 2029*

Don't miss them!

TIP

At its best oppositions — when it looks biggest and brightest — Mars is south of the celestial equator, but you can still observe it from temperate latitudes in the Northern Hemisphere.

TRACKING MARS'S BACKTRACKING

A basic project for beginning planet gazers is to track the motion of Mars across the constellations; all you need are your eyes and a sky map.

Locate Mars among the stars and mark that position with soft pencil on your map. If you repeat this observation on each clear night, you can see a pattern emerge that puzzled the ancient Greeks and led to complicated theories — most of them wrong.

Most of the time, Mars moves eastward from night to night, just as Earth's Moon moves eastward across the constellations. The Moon keeps going, but Mars sometimes reverses course. For two to almost three months (62 to 81 days) at a time, Mars heads west across the constellations, moving backward by 10° to 20°. After this period, it gets back on track and heads east again. The backtracking is called Mars's *retrograde motion*.

The backtracking isn't a case of Mars not knowing whether to come or go. The retrograde motion is just an effect produced by Earth racing around the Sun. While you chart Mars's motion, you stand on Earth, which races around the Sun once every 365 days. Mars moves slower, making one full orbit in 687 days. As a result, when we pass Mars on our inside track (lapping it), Mars seems to move backward against the reference frame of the distant stars. But in reality, Mars always forges ahead.

The easiest Martian surface feature to spot with a small telescope is usually Syrtis Major, a large dark area extending northward from the equator. Mars's day is nearly the same as Earth's: 24 hours, 37 minutes. So if you look at Mars off and on during a night, you may be able to see Syrtis Major move slowly across the planet's disk as Mars turns. Experienced amateur planetary observers may see its polar caps and other markings as well.

TIP

NASA images of Mars, taken by interplanetary probes and the Hubble Space Telescope, are much too detailed to guide you in small-telescope observation. You need a simple *albedo map*, which charts and names the bright and dark areas on Mars as visible with small telescopes. An albedo map offers more detail than the average observer ever sees, and it offers a good guide and a challenge to your observing skills. You can find such a map (prepared by the Association of Lunar and Planetary Observers) on the NASA MarsWATCH website, at `mars.jpl.nasa.gov/MPF/mpf/marswatch/marsnom.html`. If you prefer a map in color, you can purchase the Mars Topography Map, prepared for NASA by the U.S. Geological Survey, at the Shop section of the Sky & Telescope website for about $10. It has more detail than you'll ever see on Mars, unless you visit there on some future day.

Astronomers rate sky conditions in terms of *seeing* (the steadiness of the atmosphere above the telescope), *transparency* (the freedom from clouds and haze), and *sky darkness* (the freedom from interfering artificial light, moonlight, or sunlight). When observing a bright planet such as Mars, good seeing is the most important factor, and dark sky is least important. But the darker the sky, the steadier the air, and the higher the transparency, the more you can enjoy the night.

TIP

With good seeing, the stars don't twinkle quite as much, and you can use a higher-magnification eyepiece with the telescope to bring out fine details on Mars or another planet. When the seeing isn't good, the telescopic image looks blurred and seems to jump around. Under adverse conditions, high magnification is useless; you only magnify the blurred, jumping image. Use a low-power eyepiece for the best results.

Unfortunately, even when atmospheric conditions are ideal at your observing site and when an opposition of Mars is in progress, disaster may strike. Mars is a planet that experiences worldwide dust storms, which hide its surface features from view.

In the past before Mars-orbiting spacecraft were launched, professional astronomers would rely on amateur astronomers to help monitor Mars, to let them know when a dust storm began and to report other pronounced changes in the appearance of the planet.

You need experience to become a confident telescopic Mars observer. As a beginning observer, don't assume that a great dust storm is in progress just because you can't make out any detail. Get accustomed to seeing Mars on different occasions, when seeing conditions are sometimes better and sometimes worse. Only then should you consider that, when you can't see details, the planet is at fault, not the viewing circumstances or your inexperience. Remember this scientific motto: "The absence of evidence is not necessarily evidence of absence." You may not see Mars surface details the first time you look, but that doesn't mean that a dust storm is obscuring your view. As a telescopic observer, you have to train your viewing skills, just as gourmets and wine lovers train their palates. I'll drink to that.

TIP

In fact, you don't need to head for the roof or the backyard with a telescope to check out Mars. You can join planetary experts in their research by heading to NASA's Be a Martian! website (beamartian.jpl.nasa.gov). Register there, start poring over high-resolution close-ups of the Martian surface, and help the scientists characterize the fine topographic details. Citizen Science rises again!

And just for your information, Mars has only two known moons, Phobos and Deimos. These tiny celestial bodies aren't visible with small telescopes.

Outdoing Copernicus by observing Mercury

Historians say that the great Polish astronomer Nicholas Copernicus (1473–1543), who proposed the *heliocentric* (Sun-centered) *theory* of the solar system, never spotted the planet Mercury.

But Copernicus didn't have modern aids, such as smartphone apps, desktop planetariums, astronomy websites, or even monthly astronomy magazines (see Chapter 2). You can use these aids to find out when Mercury will be best placed for observation during the year: the times of greatest western and eastern elongation (terms that I cover in the earlier section "Understanding elongation, opposition, and conjunction"), which occur about six times each year. Better yet, get acquainted with the Mercury Chaser's Calculator on the website of the Association of Lunar and Planetary Observers. Surf to `alpo-astronomy.org`, click on "Mercury Section," and then click on "Mercury Chaser's Calculator." Then read or print out the details of Mercury elongations for the current year.

At temperate latitudes, such as those of the continental United States, Mercury is usually visible only in twilight. By the time the sky is dark, well after sunset, Mercury has set, too. And in the morning, you can't spot Mercury until the impending dawn starts to light the sky. Mercury resembles a bright star but appears much dimmer than Venus in the west at dusk or in the east before dawn.

Be an early riser for Mercury

Mercury is much smaller than Venus, but you can see its phases through your telescope. The best time to do this is when Mercury is at western elongation and appears in morning twilight. The atmospheric steadiness, or seeing, is almost always better low in the east near dawn than it is low in the west after sunset, so you get a much sharper view in the morning.

MERCURY LOVERS CHOOSE MORNING

Here's why seeing is better near the dawn horizon than near the sunset horizon: By sunset, the Sun has warmed Earth's surface all day, so as you look out low in the sky to the west, you look through turbulent currents of warm air rising from the surface. But in the morning, Earth has had all night to cool down and stabilize. It takes a few hours for the rising Sun to warm the land and mess up the seeing again.

TIP

You need a viewing site with a clear eastern horizon because Mercury doesn't get high in the sky when the Sun is below the horizon. If you have trouble spotting it with your naked eye, sweep around that part of the sky with a pair of low-power binoculars. And if you have a computerized telescope with a built-in database, you can just punch in "Mercury" and let the telescope do the finding.

Don't expect to see surface markings

Seeing surface markings on Mercury with a small telescope, or with almost any telescope on Earth, is extremely difficult. Mercury's apparent size at greatest elongation is only about 6 to 8 arc sec (see the sidebar "Wait just an arc minute [or second]" earlier in this chapter for details).

Some advanced amateur observers report seeing surface markings on Mercury or photographing them with digital cameras mounted on their telescopes, but I'm not aware of any useful information that has come from such sightings. A few of the greatest planetary observers in earlier times thought that they could see and draw the surface markings. From their drawings, the observers tried to deduce the rotation period, or "day," of Mercury. They concluded that Mercury's day was equal to the length of the year on Mercury, or 88 Earth days. But they were wrong. Radar measurements later proved that Mercury turns once every 59 Earth days. So a year on Mercury is less than two of its days. Go figure!

In any case, when you find out how to spot Mercury by eye and then view its phases with your telescope, you'll have outdone Copernicus!

Chapter **7**

Rock On: The Asteroid Belt and Near-Earth Objects

Asteroids are big rocks that circle the Sun. The vast majority of asteroids are safely beyond the orbit of Mars in an area called the *asteroid belt*, but thousands of other asteroids follow orbits that come close to or cross Earth's orbit. Many scientists believe that an asteroid hit Earth about 65 million years ago, wiping out the dinosaurs and many other species.

In this chapter, I introduce you to these space rocks and explain the best ways to observe them. And in case you're worried, I tell you the truth about the risk of an asteroid hitting Earth in the future and fill you in on the research scientists are conducting to deal with the possibility.

Taking a Brief Tour of the Asteroid Belt

Asteroids are also called minor planets because when they were first discovered, experts thought they were objects like the planets. But astronomers now believe that asteroids are remnants of the formation of the solar system — objects that never merged with enough additional space debris to grow into planets. Some asteroids, such as Ida, even have their own moons (see Figure 7-1). Asteroids are made of silicate rock, like the rocks of Earth, and of metal (mostly iron and nickel). Some asteroids may also contain carbonaceous (or carbon-bearing) rock. And in recent years, ice has been found on some asteroids.

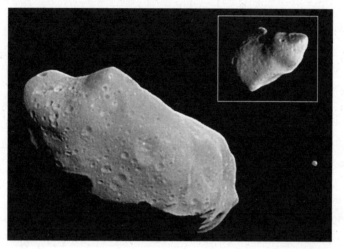

FIGURE 7-1:
The asteroid Ida has its own moon, Dactyl.

Courtesy of NASA

Most of the known asteroids are in a huge, flat region centered on the Sun and located between the orbits of Mars and Jupiter. We call this region the *asteroid belt.* Asteroids range in size from Ceres, which is 587 miles (945 kilometers) in diameter, to large meteoroids, which are just fragments of asteroids (see Chapter 4). A boulder-size space rock is either a very small asteroid or a very large meteoroid; feel free to take your pick. But Ceres is both an asteroid and a dwarf planet, according to a 2006 decision 2006. I explain the new category of dwarf planet in Chapter 9.

Asteroidal meteoroids, which I describe in Chapter 4, are made of rock and/or iron; when they fall to Earth, they're called meteorites. You can find meteorites in many natural history and geology museums. You can see the biggest meteorite ever taken to a museum (or anywhere else) at the American Museum of Natural History in New York City. The 34-ton iron mass, which is the largest piece of the Cape York meteorite from Greenland, is called Ahnighito. Check it out on the museum website at www.amnh.org/exhibitions/permanent-exhibitions/earth-and-planetary-sciences-halls/arthur-ross-hall-of-meteorites. The Hoba meteorite in Namibia is the largest one known. At about 60 tons, it just lies in place.

Table 7-1 lists the four biggest objects in the asteroid belt. The two largest, Ceres and Pallas, are at nearly the same average distance from the Sun, although Pallas has a much more elliptical orbit.

TABLE 7-1

The "Big Four" of the Asteroid Belt

Name	Diameter in Miles (Kilometers)	Length of Day in Hours	Mean Distance from the Sun (AU)
Ceres	584 (940)	9.1	2.77
Pallas	339 (545)	7.8	2.77
Vesta	326 (525)	5.3	2.36
Hygiea	253 (407)	27.6	3.14

As of early 2017, there were about 728,000 known asteroids, of which roughly 20,000 had been named. (That count includes one that the International Astronomical Union was kind enough to name after me; I'm glad they didn't just call it "Dummy.") Most were found in recent years by robotic telescopes designed for the purpose, but experienced amateur astronomers who mount digital cameras on their telescopes are also making discoveries.

You can readily see the largest asteroids, such as Ceres and Vesta, through small telescopes (see the later section "Searching for Small Points of Light" for more about observing asteroids).

Ceres and Vesta are so big that their own gravity makes them round. But small asteroids are often shaped like potatoes and frequently look blasted apart (see Figure 7-2) because, indeed, they have been. The asteroids in the belt constantly collide with each other, making impact craters and breaking off big and little chips. The big chips are simply smaller asteroids, and the little chips are asteroidal meteoroids.

At rare intervals, small asteroids (or large meteoroids) smash into Earth (see the next section for more on this phenomenon). Asteroid impacts (and comet impacts) have also covered the Moon, Mars, and Mercury with craters; Venus has craters, too, but not as many. (I describe these cratered objects in Chapters 5 and 6.)

Asteroids have craters as well, but they're much harder to see with telescopes because the asteroids themselves are so small. In most telescopes, an asteroid is just a point of light, like a star. NASA's Dawn space probe arrived at Vesta in July 2011 and studied it in detail before departing for Ceres, where it arrived in March 2015. To see pictures from the mission, visit dawn.jpl.nasa.gov.

Courtesy of NASA

The highlights of what Dawn observed at Vesta and Ceres include the following:

» Two huge impact basins (large craters) dominate Vesta's south polar region; the larger one, Rheasilvia, is 310 miles (500 kilometers) wide and about 12 miles (19 kilometers) deep, with a central peak higher than the Mauna Kea volcano in Hawaii.

» Something Dawn didn't find on Vesta is volcanic features of any kind, although it did detect surface minerals like those in Earth's basaltic lava flows.

» Vesta has *differentiated,* meaning that different layers formed inside, just as they have on Earth (see Chapter 4); it may be a surviving example of a protoplanet, meaning an early stage in the formation of a planet. If so, it's a case of arrested development.

» Ahuna Mons, a mountain on Ceres 11 miles (18 kilometers) wide and 2.5 miles (4 kilometers) high, is a *cryovolcano.* That is, it's a geological formation that channeled mineral-laden water from below to the surface of Ceres, where it froze, depositing the minerals that form the mountain.

» Unusual bright white spots appear on the floor of Ceres' Occator crater. One is a pit about 6 miles (10 kilometers) wide and 0.3 miles (0.5 kilometers) deep that contains ice that forms as liquid water rises to the surface and freezes.

When the sun warms the ice, it changes from ice to vapor and produces thin clouds in the crater that come and go with the day/night cycle. Ahuna Mons may be a dead cryovolcano, but the Occator pit is still active.

An earlier NASA probe paved the way for Dawn by exploring the 21-mile-long (34-kilometer-long) asteroid Eros, which crosses inside the orbit of Mars. The Near Earth Asteroid Rendezvous–Shoemaker probe orbited Eros for a year and then landed on February 12, 2001. You can see a video of this oblong asteroid rotating at near.jhuapl.edu; just click on "Movies," find "Eros Full-Rotation Movie," and select one of the listed movie formats to play on your computer.

More asteroid visits are in the future. NASA launched its Origins, Spectral Interpretation, Resource Identification, Security, Regolith Explorer (OSIRIS-REx) in September 2016. It will rendezvous with asteroid Bennu in 2018, map it, and try to collect a sample of loose surface material without actually landing. If all goes well, a capsule with two ounces of asteroid soil will parachute down on Utah in 2023.

Also on the agenda, pending final approval, are the Lucy (2021 launch) and Psyche (2023) spacecraft. Lucy will explore 6 of the over 6,000 *Jupiter Trojans,* meaning asteroids that are trapped in Jupiter's orbit around the Sun. (If you think of Jupiter's orbit as a circle of 360 degrees, some of the Jupiter Trojans are about 60 degrees ahead of Jupiter as it circles the Sun, and the rest are 60 degrees behind Jupiter.) Psyche will visit an asteroid named Psyche (who would have guessed?), a 130-mile (209-kilometer) diameter object astronomers believe may be almost all iron and nickel. For what the mission will cost, it may as well be gold.

Understanding the Threat That Near-Earth Objects Pose

Not all asteroids are orbiting safely beyond Mars. Thousands of small asteroids follow orbits that cross or come close to Earth's orbit. Astronomers call these neighbors *near-Earth objects (NEOs),* and they've classified 1,773 of them as *potentially hazardous asteroids (PHAs)* as of February 2017. Some day, one of these scary neighbors may come uncomfortably close to Earth or even strike our planet. The Minor Planet Center (MPC) of the International Astronomical Union keeps tabs on PHAs, and several observatories sweep the skies to discover more of them.

TIP

The MPC's website (www.minorplanetcenter.net) offers a mix of information for experts and amateur astronomers alike, including maps of the inner and outer solar system, updated daily, that show where the planets and many of the asteroids are located in space.

Astronomers don't know of any specific object that's currently a menace to Earth. But a rock a few miles in size that strikes Earth at 25,000 miles per hour (11 kilometers per second) would cause a far greater catastrophe than the simultaneous explosion of all the nuclear weapons ever made. It would be a rare case when astronomy isn't fun. Asteroids about that big collide with Earth every 10 million years, give or take, and smaller ones hit more often. Larger asteroids impact Earth much less frequently because, the bigger the asteroids, the fewer there are of them.

Conspiracy theorists think that if astronomers did know about a doomsday asteroid, we wouldn't tell. But face it, if I knew the world was in danger, I'd settle my affairs and head for the South Seas instead of sitting around finishing this chapter!

In 1998, the Hollywood movies *Armageddon* and *Deep Impact* gave sensationalized versions of what could happen if a large asteroid or comet was on a collision course for Earth. Such catastrophe stories are inspired, in part, by the widely accepted conclusion that an asteroid about 6 miles (10 kilometers) wide struck Earth about 65 million years ago. The Chicxulub crater, a 110-mile-wide (180-kilometer-wide) geologic formation resting partly on Mexico's Yucatan peninsula and partly offshore in the Gulf of Mexico, may be the surviving trace of the impact, which is theorized to have wiped out the dinosaurs. It certainly didn't do them any good.

The action of weather and geological processes, such as mountain building, erosion, flooding, and volcanism, has eroded the impact craters on Earth and destroyed most of them. As of 2017, astronomers know of 190 craters, ranging from the 44-foot-wide (30-meter-wide) Carancas crater in Peru — made by a meteorite strike in 2007 — to the 2-billion-year-old, 100-mile wide (161-kilometer-wide) Vredefort impact structure in South Africa. You can discover more about them and see pictures at the Earth Impact Database website of the University of New Brunswick (www.passc.net/EarthImpactDatabase/index.html).

Some impact craters haven't been found yet. If you do a lot of hiking, exploring, or aerial sightseeing, you may even discover a crater. Then you'll make your own impact on science.

TIP

An impact by a small asteroid caused the famous Meteor Crater (which ought to be called Meteoroid Crater or Asteroid Crater) in northern Arizona, near Flagstaff. The site is well worth a visit because it's the largest well-preserved impact crater on Earth. When I was a young scientist at the Kitt Peak National Observatory in Arizona during the 1960s, you could just tell the receptionist at Meteor Crater that you were an astronomer and get in free. Nowadays, everybody over the age of 5 pays admission, but it's worth it. Check out the website at www.meteorcrater.com for information on the hours the crater is open to visitors, directions, the Meteor Crater RV Park, and more.

For a brief time in March 1998, many people feared that a small, newly discovered NEO could strike Earth in the year 2028. Astronomers eliminated that possibility within a day when additional observations showed that the asteroid's orbit won't intersect Earth. Some experts even disagreed with the initial prediction — as experts often do.

TIP

Although Earth appears to be safe for now, scientists may discover an NEO on a collision course with Earth in the future, and they're studying the options for what can be done in such an eventuality. So many NEOs have been discovered (mostly very small) that about one a day is observed or predicted to pass by Earth. You can subscribe free to daily minor planet announcements of these events by email. They tell the name, discovery date, closest approach distance, speed, and size of each visiting asteroid. Just sign up for the bulletins at `www.minorplanetcenter.net/daily-minor-planet`.

If you want to explore the consequences of a collision, download the CraterSizeXL app for your iPhone or iPad. Insert the size and speed of a falling body, and the app calculates the size of the crater that it will make. Or you can visit the Impact Earth! website at Purdue University (`www.purdue.edu/impactearth`) and calculate the catastrophe online.

When push comes to shove: Nudging an asteroid

Some experts propose developing a powerful nuclear missile to intercept a killer asteroid before it can strike Earth. But if we blow up an asteroid heading our way, the results could be worse than the damage caused by the impact of the intact asteroid. It would be like the scene in the Disney movie *Fantasia* in which the sorcerer's apprentice chops up the out-of-control magic broom that won't stop fetching water. He ends up with a whole bunch of little brooms, each of which starts fetching water.

If we blow up an asteroid with a nuclear bomb, a swarm of smaller rocks would follow the same deadly trajectory. The rocks would pack more wallop than all the Pentagon's weapons combined. A better idea is to use the nuclear missile (or perhaps some other kind of missile) to only nudge the asteroid so it passes through its intersection point with Earth's orbit a little early or a little late, when Earth hasn't reached that spot yet or has already gone past. Phew!

The problem with nudging an asteroid is that scientists don't know how much force to apply. We don't want to break it up, but because we don't know the mechanical strength of asteroids, we don't know how hard to hit it. Asteroids may be made of strong rock or fragile stone. Some may be mostly solid metal. If we don't know our enemy, we can make things worse by striking in the wrong way.

Rather than risk unwanted consequences by trying to blow up or poke a threatening asteroid, engineers suggest using a *gravity tractor*. A massive spacecraft would fly along with the asteroid for a period of years. Without ever touching the threatening body, the spacecraft would slowly change the rate at which the asteroid is moving, thanks to the gravitational attraction between the two bodies. This plan should keep the asteroid intact but move so it just misses Earth. The problem with this method is that it involves launching a very heavy spacecraft to rendezvous with the asteroid and travel close to it for a decade or more. Not enough time may remain to do this after the collision is predicted.

Experts have proffered a lot of other ideas about how to avoid a threatening asteroid, but it's not clear which method is best. In *Fantasia*, the sorcerer himself breaks the spell on the enchanted broom, but without a sorcerer to make the asteroids disappear, we need hard information to design a system that can safely protect Earth from asteroids.

Forewarned is forearmed: Surveying NEOs to protect Earth

Astronomers have a plan to help design a system that can protect Earth from renegade asteroids (the official numbers are in kilometers, but I also give equivalent values in miles or feet):

1. **Take a census of the NEOs to make sure that we've located almost every rock that measures 1 kilometer (0.62 miles) or more in size in our region of the solar system.**

   That was the original goal, but now we're aiming to find NEOs down to 140 meters (about 460 feet) across. NEOs of this size can become PHAs if their orbits take them close to Earth.

2. **Track these NEOs and compute their orbits to determine whether any are likely to strike Earth in the foreseeable future.**

3. **Study the physical properties of asteroids to discover as much as we can about them.**

   For example, make telescopic observations to determine what kind of rock or metal they're made of.

4. **When astronomers understand the threat, an engineering team can design a space mission to counteract it.**

The Panoramic Survey Telescope and Rapid Response System (Pan-STARRS) is making a major effort in the search for previously unknown NEOs. Pan-STARRS1, the part of the system that's already in full operation, is supported by 14 scientific organizations from seven nations. It's installed atop the Haleakala volcano on Maui.

As Pan-STARRS and other facilities find NEOs and determine their orbits, experts can calculate the odds that they will strike Earth in the near future. But — and this may surprise you — nobody is in charge of reacting to an asteroid collision threat if one is predicted. Defense departments and military commands worldwide are responsible for protecting their national territories and (sometimes) the territories of allied nations. But no space agency or armed force is charged with the mission of defending Earth against a threat from space. Yes, NASA has a Planetary Defense Coordination Office, but *coordination* is the key word here. The office has an organization chart and the responsibility to notify other agencies if an asteroid threat looms. But then what? Let's hope that a protective agency is established and given the necessary power and resources by the time we all need it. Otherwise, we will truly be between a rock and a hard place. Remember the dinosaurs!

Searching for Small Points of Light

Looking for asteroids is like scanning the sky for comets (see Chapter 4), except that you look for a small point of light resembling a star rather than a fuzzy image. But unlike a star, an asteroid moves perceptibly against the background of other stars from hour to hour and from night to night.

TIP

You can easily see the largest asteroids, such as Ceres and Vesta, through small telescopes; astronomy magazines publish short articles and sky charts to guide you in advance of good viewing periods (in general, there are no best times of the day or year to view asteroids). Most good desktop planetarium programs (and similar smartphone apps) display sky maps that point out the location of bright asteroids. (See Chapter 2 for more about magazines, apps, and planetarium programs; go to Chapter 3 to find out about telescopes.)

You aren't ready to search systematically for unknown or "new" asteroids until you become a skilled amateur astronomer with a few years of experience. Advanced amateurs search for new asteroids with digital cameras on their telescopes. They collect a series of images of selected areas of the sky, generally in the direction opposite the Sun (which, of course, is below the horizon). When they see a small point of light (resembling a star) change its location, they probably see an asteroid.

The easiest asteroid-related activity for beginners to try is observing occultations. An *occultation* is a kind of eclipse that occurs when a moving body in the solar system passes in front of a star. The bodies responsible can be Earth's Moon (lunar occultation, which I describe in Chapter 5), the moons of other planets (planetary satellite occultation), asteroids (asteroidal occultation), or planets (planetary occultation). Comets and the rings of planets can also cause occultations. An occultation doesn't look like much; you just see the star disappear for a short time during the eclipse.

You can enjoy an asteroidal occultation without obtaining scientific data, but what a waste of a unique opportunity! The details of an occultation differ from place to place on Earth. For example, the same occultation may last longer as seen from one place on Earth than from another, or it may not occur at a certain location. So at some locations, you see the star being eclipsed, and at other locations, you see the star without an eclipse. From occultation data, astronomers can get a more accurate picture of a number of sky objects. For example, sometimes the occultations reveal that what seems to be an ordinary star is actually a close *binary system* (two stars in orbit around a common center of mass; see Chapter 11 for details on binary stars).

The following sections tell you how to track and time asteroidal occultations.

Helping to track an occultation

Asteroidal occultations are much trickier to observe than lunar occultations because astronomers often can't predict them with sufficient precision. Astronomers go to various spots on the predicted *occultation ground track* (a narrow band across the surface of Earth where astronomers expect the occultation to be visible — just like the path of totality in an eclipse of the Sun, which I describe in Chapters 2 and 5) and attempt to observe asteroidal occultations. But because the diameters, orbits, and shapes of most asteroids aren't known with sufficient accuracy, the predictions can't be precise. Because the occultation may be visible at some places and not others, astronomers need volunteers to monitor an asteroidal occultation at many locations. Amateur observations help determine the sizes and shapes of the asteroids involved in the occultations. You can join in, too.

TIP

The International Occultation and Timing Association (IOTA) tells all you want to know about observing occultations at occultations.org. The IOTA website is regularly updated to provide the latest predictions of occultations by asteroids and other objects, so be sure to check it often.

TIP

IOTA recommends that you begin occultation study by observing with an experienced astronomer, just to get the hang of it. Once you've got that experience under your (asteroidal occultation) belt, consider downloading the free 378-page e-book, *Chasing the Shadow: The IOTA Occultation Observer's Manual,* at the site. Just pull down the Publications menu, select IOTA Observers Handbook, and click Chasing the Shadow. When you have observations of your own to report, consult Appendix F in the handbook, where part F.2, "Asteroid Occultation Report Forms," has the necessary forms and email addresses to which they should be sent. There are addresses for occultation coordinators who will accept your reports in English, Spanish, and Portuguese, and for coordinators in Australia, Europe, Japan, New Zealand, and other countries. Occultation groundtracks can cross international borders, so it's good that occultation observers can be found all over Earth.

Timing an asteroidal occultation

To make your asteroidal observation scientifically useful, you need to time it accurately and know the exact location (latitude, longitude, and altitude) of where you are when you observe the occultation. In the past, observers figured out their locations by consulting topographic maps. Nowadays, you can use a GPS receiver or smartphone app to determine the coordinates of your observing site.

Chapter **8**

Great Balls of Gas: Jupiter and Saturn

Jupiter and Saturn, located beyond Mars and the asteroid belt, are among the best sights to see through a small telescope, and at least one is usually well placed for observations. The four largest moons of Jupiter and the famous rings of Saturn are the favorite targets when amateur astronomers give friends and family some peeks through their telescopes. And although you may not be able to tell through the telescope, the underlying science of these huge planets and their satellites is fascinating, too. In this chapter, I describe the magnificent sights you can observe through your telescope and clue you in on the basic facts about the two largest planets in our solar system.

The Pressure's On: Journeying Inside Jupiter and Saturn

Jupiter and Saturn are like hot dogs with unapproved food coloring. The meat isn't the mystery; the additives are. What you see in telescopic photographs of Jupiter and Saturn are the clouds, made of ammonia ice, water ice (like the cirrus clouds on Earth), and a compound called ammonium hydrosulfide. Water-drop clouds may be a part of the mix, too. But appearances are deceiving. These cloud materials are made of trace substances. Jupiter and Saturn are mostly hydrogen and helium, like the Sun. And despite much theorizing, scientists have no proof of what makes the Great Red Spot (GRS) on Jupiter red or what produces the other off-white tints in the clouds of the two great planets.

Jupiter and Saturn are the largest of the four gas giant planets (the others are Uranus and Neptune). Jupiter has 318 times the mass of Earth; Saturn surpasses Earth's mass by about 95 times. As a result, their gravity is enormous, and inside the planets, the weight of the overlying layers produces enormous pressure. Descending into Jupiter or Saturn is like sinking in the deep sea. The farther down you go, the higher the pressure. And unlike the sea, the temperature increases radically with the depth. Don't even think about scuba diving there.

Up at the atmospheric levels where astronomers can see, in the cloud decks, the temperatures drop to −236°F (−149°C) on Jupiter and −288°F (−178°C) on Saturn. But at great depths, the squeeze is on. By the time you reach 6,200 miles (10,000 kilometers) below the clouds on Jupiter, the pressure soars to 1 million times the barometric pressure at sea level on Earth. And the temperature equals that of the visible surface of the Sun! But Jupiter is weirder than the Sun. The density of the thick gas at this depth is much higher than at the solar surface, and the hot hydrogen gas is compressed so it behaves like a liquid metal. Swirling currents of this liquid metal hydrogen generate powerful magnetic fields on Jupiter and Saturn that reach far out into space.

Jupiter and Saturn glow intensely in infrared light, each generating almost as much energy as it gets from the Sun. (Earth, on the other hand, derives almost all its energy from the Sun.) The upward-moving heat, together with heat from the downward-shining rays of the Sun, stirs up their atmospheres and produces jet streams, hurricanes, and other kinds of atmospheric storms that continually change the appearance of these planets.

Almost a Star: Gazing at Jupiter

TECHNICAL STUFF

Jupiter has about one one-thousandth the mass of the Sun. Sometimes scientists call it "the star that failed." If only it had 80 or 90 times more mass, the temperature and pressure at its center would be so high that nuclear fusion would begin and keep going. Jupiter would start shining with its own light, making it a star!

Jupiter has a diameter of about 88,700 miles (143,000 kilometers), which is about 11 times bigger than Earth. The gas giant rotates at enormous speed, making one complete turn in only 9 hours, 55 minutes, and 30 seconds. In fact, Jupiter turns so fast that the rotation makes it bulge at the equator and flatten at the poles. With a clear look in steady air, you can detect this *oblate* shape through your telescope.

The rapid spin helps produce ever-changing bands of clouds, parallel to Jupiter's equator. What you see through your telescope when you view Jupiter is really the top of the planet's clouds. Depending on the viewing conditions, the size and quality of your telescope, and circumstances on Jupiter itself, you may see from as few as 1 to as many as 20 cloud bands (see Figure 8-1).

FIGURE 8-1:
Jupiter and its spinning-induced bands of clouds.

Courtesy of NASA

Jupiter's darker bands of clouds are called *belts;* the lighter bands are *zones.* When you look through a telescope, Jupiter looks like a round disk. Right down the center of the disk is the Equatorial Zone, flanked by the North and South Equatorial Belts (NEB and SEB). In the SEB, you may see the Great Red Spot, often the most conspicuous feature on Jupiter. This atmospheric disturbance, sometimes compared to a great hurricane, has hovered in the Jupiter atmosphere for at least 120 years. In fact, the GRS may have been spotted as early as 1664.

Jupiter is easy to find because, like Venus (see Chapter 6), it shines brighter than any star in the sky. (A small exception: When its orbit takes it to the far side of the Sun, Jupiter may be slightly fainter than the brightest star, Sirius.) If you have a computer-controlled telescope that can point to the position of the planet, sometimes you can see Jupiter in the daytime. In exceptional circumstances, you may spot Jupiter with binoculars or even with the naked eye in the daytime. A deep blue sky with little or no airborne dust helps — as do smartphone apps such as SkySafari 5 (which I describe in Chapter 2).

When you can spot Jupiter with ease, you're ready for slightly more detailed observations. I provide directions for spotting the planet's features and moons in the following sections.

Scanning for the Great Red Spot

The Great Red Spot, shown in Figure 8-2, is a storm in the South Equatorial Belt that at times has been as big as Earth and sometimes bigger. Like most of Jupiter's features, it can change from day to day. Its color can grow paler or deeper. White clouds, which are big enough to see with some amateur telescopes, form near the spot and move along the SEB. Sometimes a cloud in the SEB or another belt seems to be drawn out across the planet, stretched mostly in longitude. A cloud with this linear shape is known as a *festoon*, and spotting this interesting display is indeed a festive occasion!

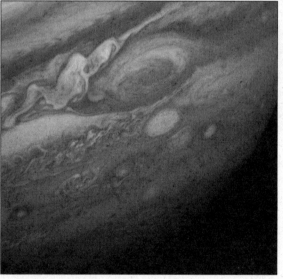

FIGURE 8-2:
Jupiter's Great Red Spot makes for stormy viewing.

Courtesy of NASA

Observations in infrared light show that the atmosphere above the Great Red Spot is much warmer than anywhere else across the planet. This discovery revealed that heat is flowing up through the GRS from below at a greater rate than in other parts of Jupiter.

JUPITER'S INVISIBLE ACCESSORIES

Jupiter has rings like Saturn's (which I describe in the section "Ringing around the planet" later in this chapter) and a magnetosphere of energetic subatomic particles, like Earth's magnetosphere (see Chapter 5). But the Jovian rings are too dark to see in amateur (and most professional) telescopes. And Jupiter's magnetosphere is much larger and more energetic than Earth's. The rings are dark because they're made of microscopic rock particles, while Saturn's rings are bright because they're mostly ice.

The Jovian magnetosphere bombards anything that moves through it, delivering a heavy radiation dose. The dose is higher for spacecraft in polar orbits, so it was only recently that NASA placed a Jupiter probe in such an orbit. The polar probe, Juno, reached the giant planet on July 4, 2016. Juno's getting great views of the polar clouds and bright auroras on Jupiter and some pictures of lower-latitude latitude regions as well. You can see photos of Jupiter, its rings, and its moons from the Galileo and Juno spacecraft and others at the Planetary Photojournal site (photojournal.jpl.nasa.gov). Just click on the image of Jupiter and follow the links.

Be alert: The GRS has been shrinking over the past 20 years. Will it vanish like Earth storms always do or reverse course and grow again? Watch it when you can. Amy Simon, NASA's Great Red Spot expert, wrote in the March 2016 *Sky & Telescope* that "amateurs' observations are critical for monitoring the GRS."

If you don't see the Great Red Spot at first, it may be in a pale condition, but it more likely has rotated to the back side of Jupiter or may have shrunk so small that you can't identify it. Wait for Jupiter to rotate back around to make sure. In the early 1990s, the SEB itself seemed to disappear overnight. Later it reappeared. This belt has faded away and then appeared again several times since then. Amateur astronomers are sometimes the first to spot SEB events like these. So while you enjoy the spectacle of Jupiter's belts and spots, be alert for something new.

Shooting for Galileo's moons

Whenever the seeing is good, your telescope will reveal structure in Jupiter's cloud tops — belts, zones, spots, and maybe more — and you may see one or more of the planet's four large moons: Io, Europa, Ganymede, and Callisto. (Check out a photo of Jupiter and these moons in the color section of this book.)

Jupiter's four prominent moons (it also has 63 smaller known moons as of February 2017) are known as the Galilean moons or Galilean satellites, named after Galileo, their discoverer. Each of the big four moons orbits almost exactly in the equatorial plane of Jupiter, so each moon is always right overhead somewhere on Jupiter's equator. Any telescope worth owning can spot the Galilean moons, and many people can even see two or three of them through a good pair of binoculars. However, Io, the innermost of the Galilean moons, is hard to spot through binoculars because it orbits very close to the bright planet.

You can't see enough detail on any of Jupiter's moons with your own telescope to figure out what their surfaces are like, but you can notice differences in their brightnesses and, with careful study, perhaps in their colors.

Each of the Galilean moons is literally a world unto itself, with composition and landscape that give it individual character. Here are basic details on the four of them:

>> **Callisto:** Callisto has a dark surface marked by many white craters. The surface is probably dirty ice — a mixture of ice and rock. The impacts of asteroids, comets, and big meteoroids have exposed the underlying clean ice — hence, the white craters. The most noticeable marking is Valhalla, a huge-ringed impact basin about as large as the continental United States (judging the size by the outermost ridge).

>> **Europa:** This moon has a ridged terrain that looks like rafts of ice. The surface is a frozen crust about 10 miles (16 kilometers) thick that lies above an underground ocean, perhaps 60 miles (97 kilometers) deep (these are rough numbers). Europa is one of six places in the solar system outside of Earth where scientists have strong evidence for subsurface liquid water. (The others are Ganymede, Callisto, Saturn's moons Titan and Enceladus, and asteroid Ceres, which I describe in Chapter 7.) Some experts think conditions are suitable for primitive life forms to exist in the Europa ocean. Planning is underway for a Europa Lander probe that would be equipped to search for such organisms.

>> **Ganymede:** At 3,270 miles (5,262 kilometers) in diameter, Ganymede is the largest moon in the solar system (even larger than 3,032-mile-wide Mercury). Ganymede's blotchy surface consists of light and dark terrains, perhaps ice and rock, respectively.

>> **Io:** This moon's surface is peppered with more than 400 volcanoes. Io is the only place other than Earth where we have definite evidence of ongoing volcanism as we know it on Earth, with hot lava emerging from underground. (But see my description of Saturn's moon Enceladus later in this chapter for a case of icy volcanism.) Io has no visible impact crater because lava from the ubiquitous volcanoes has covered up all impact sites.

Although you can't enjoy the kind of up-close and personal view achieved with sophisticated space equipment, you can observe with your telescope some interesting aspects of these moons as they orbit Jupiter. I cover phenomena that may affect your view of the moons — such as occultations, transits, and eclipses — in the following sections.

Recognizing moon movements

Io, Ganymede, Europa, and Callisto are always moving, changing their relative positions and appearing and disappearing as they revolve around Jupiter. Sometimes you can see them all, and sometimes you can't. If you can't spot one of the moons, here are a few possible explanations:

>> An *occultation* may be underway, which occurs when one of the moons passes behind the limb of Jupiter (the edge of the disk you see through your telescope).

>> The moon may be in *eclipse,* which occurs when the moon moves into Jupiter's shadow. Because Earth is often well to the side of a straight line from the Sun to Jupiter, Jupiter's shadow can extend well to its side as you see it from Earth. When you see a moon in plain view well off the limb of Jupiter and it suddenly dims and disappears, it has moved into the planet's shadow.

>> The moon may be *in transit* across the disk of Jupiter; at that time, the moon is particularly tough to see because the moons are pale in color, making them hard to spot against the cloudy atmosphere of Jupiter. In fact, a moon in transit can be much more difficult to discern than its shadow.

You can also observe a *moon shadow,* which occurs when one of the moons is sunward of Jupiter and casts a shadow on the planet. The shadow is a black spot, much darker than any cloud feature, moving across the planet. The moon that casts the spot may be in transit at the time, but this isn't always the case. When Earth is well off the Sun–Jupiter line, you may see a moon off the limb of Jupiter casting a shadow on the planet.

Timing your moon gaze

TIP

Astronomy and *Sky & Telescope* magazines print monthly charts showing the positions of the four moons with respect to the disk of Jupiter night by night. (See Chapter 2 for more about astronomy magazines.) You can tell which moon is which by comparing what you see through the telescope with the chart.

Remember the following general rules as you watch Jupiter's moons:

>> All four Galilean moons orbit around Jupiter in the same direction. When you see them on the near side, with respect to Earth, they move from east to west; when they orbit on the far side, they travel from west to east.

>> A transiting moon is moving westward, and a moon about to be occulted or eclipsed is heading eastward (following the east–west geographic directions in the sky of Earth).

Under excellent viewing conditions, you can discern one or two markings on Ganymede, the largest Galilean moon, with a 6-inch or larger telescope. (See Chapter 3 for information on telescopes.) But to see the details of the surface, you need an image from an interplanetary spacecraft that visited the Jupiter system.

COM(ET)ING WITHIN STRIKING DISTANCE

On rare occasions, a comet or an asteroid strikes Jupiter, which may cause a temporary dark blotch atop the cloud belts that can last for months. Scientists didn't know this until July 1994, when large chunks of the broken comet Shoemaker-Levy 9 struck Jupiter. Since then, astronomers have gone back through old records of the markings on Jupiter and have found some suspicious features that may have been created in the same way.

Since 1994, astronomers have known that if a dark blotch occurs on Jupiter, it may be debris from an impacting object, not just another cloud formation on the belted planet. So they've been looking for new Jupiter impacts; through the end of 2016, amateur astronomers have spotted five more. These advanced amateurs mount video cameras on their telescopes. Sometimes they detect the dark blotch on the cloud tops caused by the breakup of the striking object, sometimes they photograph the infalling object just before it hits, and occasionally they observe a bright flash from the impact.

The first discovery came in July 2009, when Anthony Wesley discovered a fresh blotch on Jupiter with his 14.5-inch telescope near Canberra, Australia. He alerted scientists, and the Hubble Space Telescope photographed the blotch, which measured 5,000 miles (8,000 kilometers) across. (The impacting body was much smaller, just as a burning house can be dwarfed by the cloud of smoke that it produces.) In June 2010, Wesley observed and recorded a brief flash of light near the edge of Jupiter's disk. It was the glow from a large meteoroid fall on Jupiter, and another amateur in the Philippines also recorded it on video. Keep your eyes (and telescope!) peeled for unusual features as you observe Jupiter. Amateurs in Ireland, Japan, and Wisconsin have also photographed Jupiter impacts.

If you spot something new and exciting on Jupiter or Saturn (which I describe in this chapter), send a report to the Planetary Virtual Observatory and Laboratory (online at pvol2.ehu.eus/pvol2), the Jupiter Section of the Association of Lunar and Planetary Observers (www.alpo-astronomy.org/jupiterblog/), or the British Astronomical Association's Jupiter Section (www.britastro.org/section_front/15). Check these websites for useful observing information and recent reports on Jupiter.

Our Main Planetary Attraction: Setting Your Sights on Saturn

Saturn is the second-largest planet in our solar system, with a diameter of about 75,000 miles (121,000 kilometers). Most people are familiar with Saturn because of its striking set of rings. For centuries, astronomers thought Saturn was the only planet that had rings. Today we know that rings encircle all four giant gas planets: Jupiter, Saturn, Uranus, and Neptune. But most of the rings are too dim to see through telescopes from the ground. The great exception is Saturn.

According to many observers, Saturn is the most beautiful planet. Not only are its famous rings easily visible through almost any telescope, but you can also spot Saturn's giant moon, Titan. Although many astronomers find Saturn's rings to be the celestial sight that most impresses their nonastronomer friends, Titan is also a worthy attraction.

In the following sections, I provide information on observing Saturn's rings, storms, and moons. Be sure to check out images of Saturn in the color section, too.

TIP

The Cassini probe is in the midst of a long tour of Saturn and its moons that is expected to continue until late 2017, when controllers will send the probe plunging into Saturn. You can see pictures and other data that Cassini obtained at the Cassini Solstice Mission page, saturn-archive.jpl.nasa.gov, and at the Cassini Imaging Central Laboratory for Operations (CICLOPS) site, www.ciclops.org.

Ringing around the planet

Saturn's rings are usually easy to see because they're large and composed of bright particles of ice — millions of little ice fragments, some larger ice balls, and maybe some pieces the size of boulders. You can enjoy the rings through a small telescope and make out their shadows on the disk of Saturn (see Figure 8-3). Under excellent viewing conditions, the *Cassini division* — a gap in the rings named for the person who first reported it — may also be discernable.

Measuring more than 124,000 miles (200,000 kilometers) across, Saturn's rings are only yards (or meters) thick. Proportionately, the rings are like "a sheet of tissue paper spread across a football field," as Professor Joseph Burns of Cornell University once wrote. But even though the rings are proportionately as thin as facial tissue, you wouldn't want to blow your nose in them. Stuffing ice up your nostrils may chill you out more than sniffing glue, but I definitely don't recommend it.

FIGURE 8-3:
Fragments of ice
and rock make up
Saturn's rings.

Courtesy of NASA

Saturn spins once every 10 hours, 32 minutes, and 45 seconds and is even more oblate — flattened at the poles — than Jupiter. The rings tend to mislead the eye a little, however, so noticing Saturn's squashed shape can be tricky.

The rings are very large but also very thin. They keep a fixed orientation, pointing face-on at one direction in space. There's a time each year when the rings are more face-on than usual, as seen from Earth, and a time three months later when they come closer to edge-on than usual.

As Saturn goes around its own 30-year orbit, sometimes the rings are precisely edge-on and seem to vanish through small (or sometimes even large) telescopes. You can't see the rings when their edges face Earth because they're extremely thin. On those occasions, with a powerful telescope, you may see the rings projected as a dark line against the disk of Saturn. The rings last disappeared in 2009 and will vanish again in 2025.

Storm chasing across Saturn

Saturn has belts and zones just like Jupiter (see the section "Almost a Star: Gazing at Jupiter" earlier in this chapter), but Saturn's have less contrast and are much harder to see. Look for them during times of good atmospheric conditions, when you can use a higher-power eyepiece on your telescope to spot planetary details.

About once every 20 to 30 years, a big white cloud, or "great white storm," appears in Saturn's Northern Hemisphere. High-speed winds spread out the cloud until it forms a thick, bright band all the way around the planet. After a few months, it may disappear. Sometimes amateur astronomers are the first to spot a new storm on Saturn. The last great white storm began in 2010, so you may have to wait a good while to see another. In the meantime, keep an eye out for smaller white clouds that can grow and spread partway around the planet.

Monitoring a moon of major proportions

Titan, Saturn's largest moon, is bigger than the planet Mercury. Its diameter is 3,200 miles (5,150 kilometers). Some large moons have thin atmospheres, but Titan has a thick, hazy atmosphere, composed of nitrogen and trace gases such as methane. Titan's atmosphere is hard to see through, but in 2004, the NASA Cassini space probe started mapping Titan's surface in infrared light (good for penetrating haze) and radar (even better). On January 14, 2005, the European Space Agency's Huygens probe landed on Titan. Most of what we know about this unusual moon comes from Cassini and Huygens.

Much of Titan's surface is relatively flat and smooth. At higher latitudes, the moon has lakes of ethane, a liquid hydrocarbon. (*Hydrocarbons* are various chemical compounds composed of hydrogen and carbon atoms; on Earth, they occur naturally in the crude oil that comes out of the ground.) One Titan lake, Ligeia Mare, is 260 miles (420 kilometers) long and up to 525 feet (8 meters) deep in places. Radar observations from Cassini recorded two echoes from the lake, one from the liquid surface and one from the lakebed, which told scientists the depth. Cassini scientist Alexander Hayes of Cornell University stated in the October 2015 *Astronomy* magazine that the liquid hydrocarbons on Titan's surface amount to 15 times more than the volume of Lake Michigan. And you don't need any fracking to pump it out (as long as you can get to Titan, of course).

On Earth, we have the *water cycle*, in which water rains down to the ground; flows into rivers, lakes, and seas; evaporates as water vapor; and rises into the atmosphere, from which it rains down again. Titan has a similar cycle, but it involves liquid hydrocarbon rain, liquid hydrocarbon lakes, and gaseous hydrocarbons. Titan is so cold that any water on the surface is permanently frozen. Some hydrocarbons also freeze: In dry spots on Titan, there are "sand" dunes, but they're not made of rock particles, like the sand dunes on Earth. Titan's dunes most likely consist of frozen hydrocarbon particles. (You've probably got some solid hydrocarbons in your home, such as polystyrene, the plastic used in disposable cups for hot coffee or tea.) A brown haze on Titan is due to airborne hydrocarbon particles. It's a natural smog.

The dunes on Titan are *aeolian*, meaning they're formed or shaped by the wind, like Earth's sand dunes in deserts or at the beach. If there were water on the surface of Titan, it would all be frozen. But in 2012, Cassini discovered a water ocean about 60 miles below the surface on Titan, where conditions are warmer. Some experts think that it may be as salty as the Dead Sea.

TIP

With a good small telescope, you can see Titan. You may be able to see two more of Saturn's moons, Rhea and Dione, when they're near their largest elongations from the planet. (See Chapter 6 for more on elongations.) You can find a monthly chart of the locations of these moons with respect to the disk of Saturn in *Sky & Telescope* magazine. Use the charts to plan your observations. As of February 2017, Saturn had 62 known moons, most of them too small to see through amateur telescopes.

TECHNICAL STUFF

BORN MOONS AND CONVERTS ORBITING IN HARMONY

Moons come in two varieties: regular and other. The regular moons all orbit in the equatorial plane of their planets, and they all orbit in the same direction in which the planets spin on their axes. This direction is called *prograde*. The regular moons almost certainly formed in place around Jupiter and Saturn, from an equatorial disk of protoplanetary and protomoon material. So Jupiter and Saturn, together with their many moons, are like miniature solar systems centered on big planets rather than stars.

But some of the small moons are like Elsa, the lioness that was "born free" and taken captive as a cub. They orbit in the direction opposite to the way the planet spins. These orbits are called *retrograde,* and they often are tilted with respect to the equatorial planes of their planets. The retrograde-orbiting moons formed elsewhere in the solar system, perhaps as asteroids, and were captured by the gravity of Jupiter and Saturn.

Jupiter has 67 confirmed moons, and Saturn has 62, as of February 2017. Each planet probably has quite a few more small ones, and astronomers keep finding more of them. Any number of known moons that you find in a printed book may be obsolete by the time you read it. Sometimes astronomers announce moons but don't count them. The officials at the International Astronomical Union want to be sure that the discoveries are confirmed. You can check the latest information on the natural satellites of Jupiter, Saturn, and other planets at the NASA Solar System Dynamics site, ssd.jpl. nasa.gov/?sat_discovery. The moons without names are provisional discoveries awaiting confirmation.

Venting about geysers on Enceladus

One of Cassini's most interesting discoveries was the existence of vents in the south polar region of Saturn's satellite Enceladus. Water vapor, ice particles, and other substances are pouring out of the vents, like cold versions of the geysers of hot steam in Yellowstone National Park. Astronomers counted 101 geysers on Enceladus, much fewer than in Yellowstone. The eruption of ice-cold material like this is called *cryovolcanism*. (I explain what a cryovolcano is in Chapter 7.) Astronomers concluded that the Enceladus geysers are fed from a subsurface body of liquid water on the moon that's warm enough to support life, if anything lives there. Fresh ice crystals from the geysers coat the surface of Enceladus, making it especially bright. Some ice particles are propelled out into space, where they merge into one of Saturn's rings.

Chapter **9**

Far Out! Uranus, Neptune, Pluto, and Beyond

Although Mars and Venus are closer to Earth, and Jupiter and Saturn are the bright, showy planets, observing the outer planets has its own mystique and offers its own rewards. This chapter introduces you to our solar system's two outermost planets — Uranus and Neptune — and describes Pluto (now called a dwarf planet). I also provide details on the moons of Uranus, Neptune, and Pluto; offer useful tips for viewing these far-out worlds; and share details about the Kuiper Belt.

Breaking the Ice with Uranus and Neptune

The following are the most important facts about Uranus (pronounced *yoo-RAN-us*, or more commonly *YOO-rin-us*) and Neptune:

» They have a similar size, with similar chemical compositions.

» They're smaller and denser than Jupiter and Saturn.

>> Each planet is the center of a miniature system of moons and rings.

>> Each planet shows signs of a long-ago encounter with a large body.

The atmospheres of Uranus and Neptune, like those of Jupiter and Saturn (see Chapter 8), are mostly hydrogen and helium. They too are gas giants, but a bit smaller. Astronomers also call Uranus and Neptune *ice giants* because their atmospheres surround cores of rock and water. The water is so deep inside Uranus and Neptune and is under such high pressure that it's a hot liquid. But when each of these planets coalesced from smaller bodies billions of years ago, the water was frozen.

Uranus has about 14.5 times the mass of Earth, and Neptune equals 17.2 Earths, but they appear nearly the same size. The lighter Uranus is a bit larger, measuring 31,770 miles (51,118 kilometers) across the equator. Neptune's equatorial diameter is 30,775 miles (49,528 kilometers).

One day on Uranus lasts about 17 hours and 14 minutes; a day on Neptune spans 16 hours and 7 minutes. So as with Jupiter and Saturn, these planets both rotate faster than Earth. Although the days on Uranus and Neptune are shorter than on Earth, the years are much longer. Uranus takes about 84 Earth years to make one trip around the Sun, and Neptune takes about 165 Earth years.

I cover more interesting facts about each planet in the following sections. Be sure to check out photos of Uranus and Neptune in the color section.

Bull's-eye! Tilted Uranus and its features

The evidence that Uranus suffered a major collision or gravitational encounter is that the planet seems to have flipped on its side. Instead of the equator being roughly parallel to the plane of Uranus's orbit around the Sun, it nearly forms a right angle with that plane so, in terms of Earth's directions, its equator runs almost north–south.

Sometimes the North Pole of Uranus points toward the Sun and Earth, and sometimes the South Pole faces our way. For about a quarter of Uranus's 84-year orbit around the Sun, its North Pole faces roughly sunward; for about another quarter, the South Pole faces roughly sunward; and the rest of the time, the Sun illuminates the whole range of latitudes from pole to pole. In 2007, the Sun was overhead at the equator on Uranus. That period would have been a good time to go to the beach, if Uranus had a beach. On Earth, the Sun is never high in the sky at the North or South Pole, but in 2028, the Sun will be high over the North Pole on Uranus.

Observations with the Hubble Space Telescope and earlier, with the Voyager 2 space probe, show that Uranus has a changing pattern of cloud belts. In 2006, a large dark spot appeared and in 2011 Hubble photographed an aurora on Uranus, the first aurora seen since the Voyager 2 flyby in 1986. The changing cloud patterns on Uranus may be related to the seasons on that planet.

As of February 2017, Uranus has 27 known moons. It also has a set of rings. The rings are made of very dark material, probably carbon-rich rock, like certain meteorites known as carbonaceous chondrites. The moons and rings of Uranus orbit in the plane of its equator, just as the Galilean moons orbit in the equatorial plane of Jupiter (see Chapter 8), so the rings and the orbits of the moons of Uranus are at nearly right angles to the plane of the planet's orbit around the Sun.

You can think of Uranus and its satellites as a big bull's-eye that sometimes faces Earth and sometimes doesn't. One or more large objects probably struck Uranus long ago and tilted it from its natural position. That's no bull.

Against the grain: Neptune and its biggest moon

Neptune's axis is tilted 28° from the perpendicular to its orbit plane, a bit more than Earth's tilt of 23.5°, which I described in Chapter 5. Its rings are very dark, like those of Uranus, and probably consist of carbon-bearing rock.

Neptune has 14 known moons as of February 2017. Its largest moon, Triton (which is larger than Pluto), has a diameter of 1,682 miles (2,707 kilometers). Seen from north and above, Neptune, like all the planets in our solar system, revolves counterclockwise around the Sun. Most moons revolve counterclockwise around their planets. But Triton, which resembles a cantaloupe in photos from Voyager 2, goes against the grain, traveling clockwise around Neptune. (In other words, it has a retrograde orbit, a term that I define in Chapter 8.) After mulling it over, scientists concluded that Neptune captured Triton early in solar system history. Expert opinions vary, but a leading theory is that, back then, Neptune had a near collision with a binary system of two small objects from the Kuiper Belt (which I discuss at the end of this chapter). Neptune grabbed Triton, which was one half of the binary, and the other little planet flew away. Astronomers need more facts to be able to test this theory.

Triton consists of ice and rock, so it seems more like Pluto (see the next section) than Uranus or Neptune. Its surface is shaped by eruptions and flows of cold icy substances rather than hot, molten rock. (It's the result of *cryovolcanism*, which I describe in Chapter 7.) Water ice, dry ice, frozen methane, frozen carbon monoxide, and even frozen nitrogen are all present on Triton. The moon doesn't have many impact craters, probably because they got sloshed full of ice over time.

Environmental groups say that excessive tourism endangers national parks, so consider a trip to Triton instead. Its landscape is just as bizarre, and maybe as beautiful, as Yellowstone's. But if you head for Triton, expect a winter wonderland. The surface has cold surges rather than hot springs, and its geysers spew long plumes of frigid vapor rather than torrid jets of steam. Just bring a space suit and some warm booties.

Neptune's atmosphere features cloud belts, and occasionally a so-called Great Dark Spot appears, which may be a very large storm, similar to the Great Red Spot on Jupiter (see Chapter 8). Jupiter's big spot appears, fades, and reappears again, all in about the same place in the same cloud belt. But Neptune's Great Dark Spot was first found in the planet's Southern Hemisphere in 1989; then it disappeared, and later a so-called Great Northern Dark Spot came into view in the opposite hemisphere. In 2016, the Hubble Space Telescope confirmed the presence of the first dark spot on Neptune seen in the 21st century. It was in the south.

Meeting Pluto, the Amazing Dwarf Planet

For decades, astronomers deemed Pluto the most distant planet from the Sun in our solar system (see Figure 9-1). Actually, Pluto moves inside the orbit of Neptune every 248 years for a few decades at a time, but the last such inside move ended in early 1999. It won't happen again in the lifetime of anyone now inhabiting Earth, unless medical research makes major strides long before the 23rd century. But Pluto was demoted. On August 24, 2006, the International Astronomical Union (IAU) voted to reassign it to the status of *dwarf planet.*

Dwarf planets are a newly recognized class of astronomical objects, defined by the IAU as bodies with these characteristics:

>> Orbit the Sun directly (as opposed to orbiting another body, such as a planet)

>> Are massive enough that their own gravity makes them round

>> Haven't "cleared the neighbourhood" around their orbits

I surrounded the third criterion in quotation marks because that's what the IAU said, but many astronomers don't think there's a clear explanation of what "cleared the neighbourhood" means. The general idea is that a planet's gravity disturbs objects that are in nearby orbits (but not moons of the planet) so the objects — such as asteroids and comets — are thrown into orbits that take them away from the general vicinity of the planet. Many Kuiper Belt objects (see the final part of this chapter) are in Pluto's general vicinity, so Pluto supposedly hasn't cleared its area. But thousands of so-called *Trojan asteroids* are right in

Jupiter's orbit (as I describe in Chapter 7), and no one denies planet status to Jupiter. Did the IAU pick on Pluto because it's too small to defend itself? For more about this dispute, I immodestly recommend *Pluto Confidential: An Insider Account of the Ongoing Battles over the Status of Pluto*, by Laurence A. Marschall and me (BenBella Books). Pluto is so far away that scientists had only a rough idea of its geography before July 2015, when the New Horizons probe flew right by. Pluto's elongated elliptical orbit brings it within about 29.7 AU, or 2.8 billion miles (4.4 billion kilometers), of the Sun and takes it as far out as 49.5 AU, or 4.6 billion miles (7.4 billion kilometers). It may only be a dwarf, but this planet is far out!

FIGURE 9-1:
Pluto is mysterious, rocky, and icy.

Courtesy of NASA

Getting to the heart of Pluto

New Horizons's instruments photographed Pluto, scanned it to determine chemical composition, found over 20 haze layers stacked in its atmosphere, studied the known moons, and searched for more moons and for evidence of a ring around the dwarf planet. It obtained so many images and measurements that it could promptly radio only a modest selection to Earth. The rest were stored on board the probe and beamed home over the course of more than a year while the probe left Pluto far behind.

One part of Pluto struck the imagination of all who saw the first good pictures: Tombaugh Regio, informally called "the Heart." It's a huge region shaped roughly like a heart-shaped box of Valentine's Day candy and named for the late American astronomer Clyde Tombaugh, who discovered Pluto. The two lobes of the Heart are distinctly different. The west lobe, Sputnik Planitia, is bright and exceptionally smooth, while the eastern lobe is noticeably darker and much rougher.

Counting craters

One of the first things planetary scientists do when they study the solid surface of a planet (or moon or dwarf planet) is estimate how old the various parts of the surface are. Of course, the whole body formed at one time, but some surface areas were altered later as erosion or lava flows (for example) wiped out their old surfaces and produced new ones. On a cold world like Pluto, ice may have filled a crater so you no longer see it — what's left is a fresh, young surface.

The main way astronomers date solid surfaces is by counting craters. Over millions or billions of years, impacts on a planet make craters in roughly equal amounts on two areas of the same size. So if one area has significantly fewer craters than another, the region with fewer craters has been resurfaced more recently and therefore is younger. As soon as New Horizons started sending back good images, the crater count began.

Here's what the Pluto counts revealed:

>> Sputnik Planitia has no craters, at least none large enough for the resolution of the photos to reveal. So Sputnik is very young by astronomy standards — less than 10 million years old, and maybe even less than a million years.

>> The eastern lobe of Tombaugh Regio is visibly cratered; much of it is about 1 billion years old.

>> The heavily cratered, and in some places mountainous, regions of Pluto away from the Heart have ages around 4 billion years.

Examining Sputnik Planitia

Every part of Pluto seems to harbor something new and interesting for scientists, but Sputnik Planitia is the most important feature, and it's vital to understanding what's happening on Pluto. Consider these important findings on Sputnik Planitia:

>> It's an ancient *impact basin* (very large crater) about 650 miles (1,050 kilometers) across.

>> It's filled with ice — mostly frozen nitrogen — which makes it smoother than the other parts of Pluto, which have much older surfaces.

>> Large areas of its ice are arranged in polygonal cells shaped like the patterns of cracks you see on Earth where mud or clay soils have dried out.

>> Ice from highlands bordering Sputnik flows down into the basin.

>> It's on the side of Pluto that always faces exactly away from the big moon Charon; a line from the center of Charon through the center of Pluto would come out of Pluto right at Sputnik.

Here's what this all may mean:

>> After a huge impact struck Pluto long ago, the basin it made filled with nitrogen ice that (under the conditions on Pluto) is heavier than water ice but less stiff, so it flows readily.

>> The great ice mass in the basin caused Pluto to lean over until it reached a stable position in which Sputnik Planitia is right on the axis connecting Pluto and Charon.

>> The polygons on Sputnik's ice surface are convection cells, meaning heat from Pluto's interior warms blobs of frozen nitrogen underneath. They rise like bubbles of boiling water in a tea kettle, cooling and spreading out as they reach the surface. As the nitrogen cools and gets denser, it drops back down at the edges of the cells.

>> If a meteoroid falls on Sputnik, making a crater, ice flows in and fills it. After a while, no crater is visible.

>> Seasons change at Sputnik Planitia. When it's warmer, nitrogen ice vaporizes and joins the atmosphere, thickening the atmosphere. That same nitrogen gas will condense and fall like snow when it reaches places on the planet that are seasonally cold. So Sputnik is a great reservoir of nitrogen ice that periodically resupplies the atmosphere, which is itself mostly nitrogen, and Sputnik is periodically replenished with fresh ice.

Looking at Pluto's makeup

If I tell much more about little Pluto, the giant planets will be jealous of all the attention. So I'll just mention a few more interesting facts:

>> Pluto is mostly rock, but the rock is all far below the surface. The surface is a solid shell of water ice. The bedrock on Pluto's surface isn't granite; it's frozen water! Even mountains are made of water ice. The surface is so cold that water never melts or vaporizes. Other ices, such as frozen nitrogen and methane, coat the surface in places, concealing the water ice just below.

>> Between the rock interior and the ice shell surface layer is probably an underground ocean.

>> Some dark areas on Pluto, such as a long reddish band parallel to and a bit south of the equator, are coated with *tholins,* or chemical substances produced when methane particles or other hydrocarbons are struck by ultraviolet light from the sun and by galactic cosmic rays (high-energy subatomic particles from the Milky Way). The tholins drift down from the atmosphere like soot from a chimney, and they darken and redden the places where they fall.

>> Charon, the dwarf planet's moon, has a large red-brown polar cap; it's probably produced when gas that escapes from Pluto's atmosphere falls down on the big moon and is irradiated by ultraviolet light and cosmic rays, making tholins as on Pluto. Charon has no atmosphere of its own.

>> Pluto has no detectable ring and no additional moons that New Horizons could find. You can't win them all.

The moon chip doesn't float far from the planet

Pluto, like Uranus, is tilted on its side. Its equator is tilted about 120° from the plane of its orbit. Astronomers assume that Pluto, like Uranus, suffered a major collision. The colliding body probably came from the Kuiper Belt, and it may have been Charon itself.

Pluto is 1,475 miles (2,375 kilometers) in diameter and thus slightly less than twice as big as Charon, which is 753 miles (1,210 kilometers) across. They're close enough in size compared to any other planet and its largest moon that in the past they were sometimes described as a double planet.

Using the Hubble Space Telescope, astronomers found four small moons of Pluto before New Horizons arrived. They all orbit the dwarf planet in the same plane as Charon and probably were formed from the same collision.

Pluto takes 6 days, 9 hours, and 18 minutes to turn once on its axis, and Charon orbits once around the planet in exactly the same amount of time. So the same hemispheres of Pluto and Charon always face each other. In the Earth–Moon system, one hemisphere of the Moon always faces Earth, but not vice versa. Someone standing on the near side of the Moon can see our whole planet over the course of one Earth day, but a person standing on Charon always sees the same half of Pluto.

You can see a lot more about Pluto and its moons and enjoy the wonderful pictures of Pluto on the New Horizons website at The Johns Hopkins University Applied Physics Laboratory (`pluto.jhuapl.edu/`).

Buckling Down to the Kuiper Belt

Scientists estimate that about 100,000 icy bodies — called *Kuiper Belt objects* (KBOs) — larger than 60 miles (100 kilometers) in diameter orbit between Neptune's orbit and a distance of 50 AU from the Sun. That region is called the Kuiper Belt after the astronomer Gerard P. Kuiper. Almost all the KBOs are beyond the reach of backyard telescopes, unless your backyard is the site of the Palomar Observatory or a similar facility. (Amateurs with fairly large telescopes can see Pluto, which is now recognized as the first known KBO as well as being a dwarf planet.) Astronomers David Jewitt and Jane Luu discovered the first KBO other than Pluto in 1992. Since then, researchers have found more than a thousand others.

Among the many KBOs that astronomers have discovered since 1992, a few, such as Eris, rival Pluto in size. Eris is much farther from the sun than Pluto is, and it too is a dwarf planet, with at least one moon, Dysnomia.

Many of the of known KBOs share three properties with Pluto:

>> They have highly elliptical orbits.

>> Their orbital planes are tilted by a significant angle with respect to the plane of Earth's orbit.

>> They make two complete orbits around the Sun in approximately the same time that Neptune takes to make three orbits (496 years for Pluto's two orbits and 491 years for Neptune's three). This effect is called a *resonance,* and it works to keep Pluto and Neptune from ever colliding , although their orbits cross. Whew!

Pluto is safe from disturbance by the powerful gravity of the much larger Neptune, and so are the KBOs that share these three properties — called *Plutinos,* meaning "little Plutos."

Other kinds of objects that may be different from KBOs are orbiting beyond Neptune and Pluto. One such *trans-Neptunian* object, Sedna, was discovered in March 2004 at a distance of 90 AU from the Sun, well beyond the 50-AU distance where the Kuiper Belt peters out. Sedna is about 618 miles (995 kilometers) in diameter, probably large enough to count as a dwarf planet. Some astronomers believe that Sedna is a member of the Oort Cloud, a huge collection of distant comets that I describe in Chapter 4. The only *known* large planets (not dwarf planets) beyond Neptune are the planets of other stars (see Chapter 14). But check out the final section of this chapter for more on the search for new planets in our solar system!

The New Horizons probe is now far beyond Pluto on a course through the Kuiper Belt, where it will visit 2014 MU_{69}, a roughly 20- or 30- mile (32- or 48-kilometer) diameter KBO. It should get there on January 1, 2019 and ring in the New Year with new discoveries — maybe even a new moon.

Viewing the Outer Planets

With experience, you can locate the outer planets Uranus and Neptune, but tiny Pluto may be beyond your visual reach. The first time you look for any of these distant objects, seek the aid of a more experienced amateur astronomer unless you have a telescope with computerized pointing. (I recommend such telescopes in Chapter 3.) Even then, you will be better off with a little help from your friends.

Sighting Uranus

Uranus was discovered with a telescope, and it sometimes shines bright enough to be barely visible to the eye under excellent viewing conditions. When you become an experienced observer, you'll probably be able to spot it with binoculars. Through your telescope, you can distinguish Uranus from a star, thanks to

>> Its small disk, a few seconds of arc in diameter (defined in Chapter 6)

>> Its slow motion across the background of faint stars

The disk of Uranus has a pale green tint; you can make out the disk with a high-power eyepiece when viewing conditions are good. (Chapter 3 has more about telescopes and eyepieces.) You can detect the motion of Uranus by making a sketch of its relative position among the stars in the field of view. For this purpose, use a low-power eyepiece so the field of view is larger and more stars are visible. Look again in a few hours or on the following night, and sketch again.

You may glimpse a few of the biggest of Uranus's 27 known moons with large amateur telescopes, but they're better suited for study with powerful observatory telescopes. Uranus's dark rings are detectable with the Hubble Space Telescope and in images made with very large telescopes on Earth, but you can't see them with amateur instruments.

TIP

You can see the Hubble Space Telescope images of Uranus and its rings at hubblesite.org/images/news/86-uranus. You can browse images of Uranus, its moons, and the rings from the Voyager 2 space probe at photojournal.jpl.nasa.gov, the Planetary Photojournal website; just click on the labeled picture of Uranus there. (Voyager 2 is the only spacecraft that visited Uranus.)

Distinguishing Neptune from a star

Neptune appears fainter in the sky than Uranus, but it gets as bright as 8th magnitude (Chapter 1 has more about magnitudes). If Uranus challenges your observing skills, you really have to take them to the next level for Neptune when it's not at its brightest.

Neptune is about the same actual size as Uranus, but it orbits much farther away, so through a telescope, its apparent disk is smaller. You may need a large amateur telescope to discern it from a star. If you become good at perceiving pale hues in dim objects seen through a telescope, you'll be able to tell that Neptune has a blue tint.

Because Neptune orbits farther from the Sun than Uranus, it moves at a slower speed. The slower speed combined with a greater distance from Earth means that the angular rate of speed across the sky — in arc seconds per day (see Chapter 6) — is *usually* less for Neptune than for Uranus. So you may have to wait another night or two to be sure you've seen Neptune moving across the background stars.

I say "usually" because both Uranus and Neptune, like all the planets beyond Earth's orbit, show retrograde motion at times (see Chapter 6), so they seem to slow down and reverse direction now and then. If you happen to catch Uranus when it changes direction in the sky, its apparent motion is much slower than usual; by comparison, Neptune may go full tilt at that time.

The largest of Neptune's 14 known moons is Triton (see the earlier section "Against the grain: Neptune and its biggest moon" for more about Triton). After you master locating Neptune, look for Triton with a telescope of 6 inches or more in diameter on a clear, dark night. It has a large orbit and will be about 8 to 17 arc sec from Neptune (about four to eight Neptune diameters), so you may mistake Triton for a star. But by sketching Neptune and the faint "stars" around it on successive nights, you can deduce which "star" moves with Neptune across the starry background as it also moves around Neptune. That will be Triton. It takes Triton almost six days to make one full orbit around the planet.

TIP

You can browse images of Neptune and its moons from the Voyager 2 space probe at photojournal.jpl.nasa.gov, where you click on Neptune. You can see Hubble Space Telescope images at hubblesite.org/images/news/69-neptune.

Straining to see Pluto

Pluto is a much tougher viewing challenge than any planet in the solar system. It orbits far away and is small. Typically, Pluto is 14th magnitude (see Chapter 1 to find out about magnitudes). It's currently moving away from the Sun and Earth and will continue to move away for many years as it traverses its 248-year orbit.

TIP

Skilled amateurs claim to have seen Pluto with 6-inch telescopes. I recommend that you use at least an 8-inch telescope.

Pluto's largest moon, Charon, orbits very close to Pluto and revolves around it in just 6 days, 9 hours, and 18 minutes. You can distinguish it only with powerful observatory telescopes. As for viewing the smaller moons, as we say in Brooklyn, fuhgeddaboudit!

Hunting New Planet Number Nine

In January 2016, astronomers at the California Institute of Technology announced that they suspected a large planet exists far beyond Neptune and Pluto. They deduced this by comparing the orbits of Sedna (which I describe in the earlier section "Buckling Down to the Kuiper Belt") and several other trans-Neptunian objects, whose long elliptical orbits seem to line up in an unusual way. The planet, if it's real, would be massive enough to affect the orbits of these KBOs, making them line up and pulling the *perihelion points* (places of closest approach to the sun) outward from the Kuiper Belt. As this book goes to press, rival teams of astronomers are searching with very large observatory telescopes in the general area where the planet is suspected, perhaps in or near the constellation Orion.

If the estimates are accurate, Planet Nine (it won't have an official name until discovered and confirmed) follows a huge orbit, taking roughly 15,000 years to go once around the Sun. (Compare with Pluto, which takes a mere 248 years.) It may have ten times the mass of Earth, making it three-quarters the mass of the ice giant Uranus. Quite a find if true.

YOU CAN HELP FIND PLANET NINE

You may think that you have to leave the fun of finding Planet Nine up to the professional astronomers and their huge telescopes. But you would be wrong. In February 2017, NASA announced a project, Backyard Worlds: Planet 9, to enlist amateur astronomers and other citizen scientists to help in the search. That's because some experts think that an existing huge database of infrared images from the Wide-field Infrared Survey Explorer (WISE) satellite may contain unrecognized images of the suspected planet. WISE images of the same part of the sky, taken years apart, may reveal Planet 9 as it moved slowly across the background of stars. The planet should have a source of internal heat, like Jupiter and Saturn (see Chapter 8), which would show up in infrared light more readily than the planet's reflected sunlight would in visible light.

The problem is that there are huge numbers of images to search for faint signals that computer programs may miss or confuse with defects in the data. NASA needs human eyes and brains — and lots of them — to search the images. If NASA put that many employees on the project, no one would be left to launch the rockets, so it's requesting the public help out. All you need is your computer or your smartphone.

Go to www.backyardworlds.org, click on "Learn more" to watch the informative video, and, if it interests you, sign up. NASA's Marc Kuchner, the lead scientist of Backyard Worlds, says, "There are just over four light-years between Neptune and Proxima Centauri, the nearest star, and much of this vast territory is unexplored." I say, won't you help explore it? (For more about Proxima Centauri, see Chapters 11 and 14).

Even if you miss seeing Planet Nine, you may discover a new brown dwarf by aiding Dr. Kuchner's project. *Brown dwarfs* are dim purple objects intermediate between gas giant planets and real stars; they glow in infrared light, and astronomers want to find more of them. But don't ask me why purple objects are called "brown dwarfs." Maybe it's illogical, but that's how it is. (For more about brown dwarfs, see Chapter 11).

Comet McNaught sets behind Mount Paranal, in Chile, January 2007. Chapter 4 gives you the lowdown on comets, which are great blobs of ice and dust.

A view of Earth as seen from the Moon. Turn to Chapter 5 to find out more about our unique home in the solar system.

An enhanced view of the Moon, photographed by the Galileo spacecraft on its way to Jupiter. The distinct bright ray crater at the bottom of the image is the Tycho impact basin. The dark areas are impact basins filled with lava rock. Chapter 5 explores the Moon in detail.

Courtesy of NASA

Mercury, the planet closest to the Sun, is frequently invisible to the naked eye, lost in the sun's glare. You can discover more about Mercury in Chapter 6.

Courtesy of NASA/JPL

Venus is covered by clouds, but radar penetrates them to make this map of the planet. For more information about this dry, acidic planet, see Chapter 6.

A selfie by NASA's Curiosity Mars rover on lower Mt. Sharp. See Chapter 6 for details on Mars.

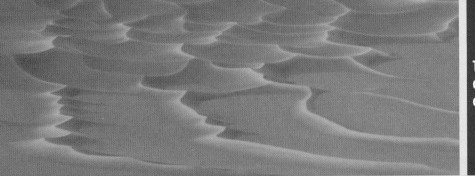

These sand dunes are in a crater on Mars.

Mars is likely to be the first planet visited by people from Earth.

Jupiter's four largest moons — Io, Europa, Ganymede, and Callisto — appear over the Great Red Spot in this montage. Chapter 8 probes this giant, gaseous planet and its moons.

Saturn's rings were photographed by the Hubble Space Telescope. You can see them with a small amateur telescope. Find out more about Saturn and its rings in Chapter 8.

Saturn's rings are shown here in false colors that reveal their nature. Those in turquoise are mostly ice; those in red contain many rock particles as well. The Cassini spacecraft photographed the rings in ultraviolet light, which is invisible to the human eye.

Like Saturn, Uranus has rings, but they can't be seen with a home telescope. Chapter 9 travels to the far reaches of our solar system to study Uranus.

Streaky white clouds and a large dark spot marked Neptune's atmosphere when this picture was taken. See Chapter 9 for details on Neptune.

Pluto, now called a dwarf planet, has a remarkable variety of surface features. To find out more about Pluto's nature, check out Chapter 9.

The Dawn spacecraft took this photograph of the largest asteroid, Ceres. Chapter 7 takes you on a tour of the Asteroid Belt.

The Jewel Box, an open star cluster of the Milky Way galaxy, is visible to the naked eye in the southern constellation Crux. For details on the star clusters and the Milky Way, see Chapter 12.

The globular cluster Messier 80 is a great ball of hundreds of thousands of stars. Chapter 12 identifies the best globular clusters to view with binoculars or a small telescope.

The Eagle Nebula is a region in the Milky Way where new stars are being born. See Chapter 12 for details.

Two spiral galaxies are caught in a cosmic collision by the Hubble Space Telescope. Chapter 12 explores the structure of spiral galaxies.

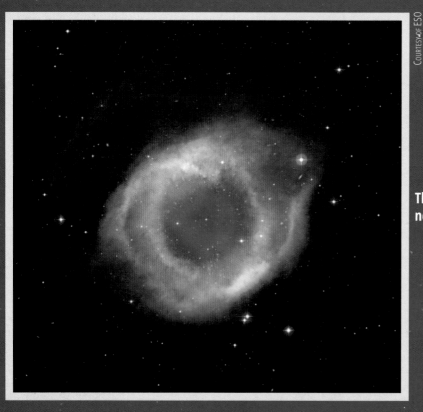

The Helix Nebula, a planetary nebula in Aquarius.

3

Meeting Old Sol and Other Stars

Chapter **10**

The Sun: Star of Earth

Although many people are attracted to astronomy by the beauty of a moonlit night and a starry sky, you need nothing more than a sunny day to experience the full impact of an astronomical object firsthand. The Sun is the nearest star to Earth and provides the energy that makes life possible.

The Sun is so familiar in daily life that people take it for granted. You may worry about getting sunburned, but you probably seldom think of the Sun as a primary source of information about the nature of the universe. In fact, the Sun is one of the most interesting and satisfying astronomical objects to study, whether with backyard telescopes or advanced observatories and instruments in space. The Sun changes day by day, hour by hour, and moment by moment. And you can show it off to the kids without keeping them up past their bedtimes.

WARNING

But don't even think about looking at the Sun, let alone showing it to a child or anyone else, without taking the proper precautions that I explain in this chapter. You don't want a view of the Sun to cost someone's eyesight. Make safety in viewing your prime consideration; when you know how to protect your vision with the proper equipment and procedures, you can follow the Sun not only daily but also over the 11-year sunspot cycle I describe later in this chapter.

This chapter introduces you to the science of the Sun, the Sun's effects on Earth and on industry, and safe solar observing. Get ready to look at the Sun in a new way — safely and with awe.

Surveying the Sunscape

The Sun is a *star*, a hot ball of gas shining under its own power with energy from *nuclear fusion*, the process by which the nuclei of simple elements combine into more complex ones. The energy produced by fusion inside the Sun powers not only the Sun itself but also much of the activity in the system of planets and planetary debris that surrounds the Sun — the solar system of which Earth is a part (see Figure 10-1, which isn't to scale).

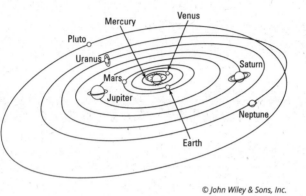

FIGURE 10-1:
Pluto and the planets orbit the Sun as part of our solar system.

The Sun produces energy at an enormous rate, equivalent to the explosion of 92 billion 1-megaton nuclear bombs every second. The energy comes from the consumption of fuel. If the Sun consisted of burning coal, it would burn up every last lump of itself in just 4,600 years. But fossil evidence on Earth shows that the Sun has been shining for more than 3 billion years, and astronomers are certain that it has been beaming away for longer than that. The estimated age of the Sun is 4.6 billion years, and it still burns strong today.

Only nuclear fusion can produce the Sun's huge total energy release — its *luminosity* — and keep it going for billions of years. Near the center of the Sun, the enormous pressure and the central temperature of almost 16 million degrees Celsius (29 million degrees Fahrenheit) cause hydrogen atoms to fuse into helium, a process that releases the great torrent of energy that drives the Sun.

About 700 million tons of hydrogen turn into helium every second near the center of the Sun, and 5 million tons vanish and turn into pure energy.

If humans could generate energy through fusion on Earth, all the problems with fossil fuel, including air pollution and the consumption of nonrenewable resources, would be solved. But despite decades of research, scientists still can't do what the Sun does naturally. Clearly, the Sun deserves further study.

The Sun's size and shape: A great bundle of gas

When I taught Astronomy 101, I'd always pose the question, "Why is the Sun the size that it is?" Hundreds of mouths would drop open and dozens of pairs of eyes would wander the room, but hardly anyone ever had a clue. It doesn't even seem like a logical question. Everything has a size, right? So what?

But if the Sun is made of nothing but hot gas (and it is), what keeps it together? Why doesn't it all blow away, like a puffed-out smoke ring? The answer, my friend, is that gravity keeps the Sun from blowing in the wind. Gravity is the force, which I describe in Chapter 1, that affects everything in the universe. The Sun is so massive — 330,000 times the mass of planet Earth — that its powerful gravity can hold all the hot gas together.

Well, you may be wondering, if the Sun's gravity pulls all its gas together, why isn't it squeezed down into a much smaller ball? The answer is the same thing that sells many a used car: high pressure. The hotter the gas, and the more gravity (or any other force) that squeezes it together, the higher its pressure. And gas pressure inflates the Sun just like air pressure inflates an automobile tire.

Gravity pulls in; pressure pushes out. At a certain diameter, the two opposite effects are equal and in balance, maintaining a uniform size. The certain diameter is about 864,500 miles (1,392,000 kilometers), or about 109 times Earth's diameter. You can fit 1,300,000 Earths inside the Sun, but I don't know where you'd get them.

The Sun is round for much the same reason: Gravity pulls equally in all directions toward the center, and pressure pushes out equally in all directions. If the Sun rotated rapidly, it would bulge a little at the equator and flatten slightly at the poles due to the effect that people often call *centrifugal force.* But the Sun rotates at a very slow rate — only once every 25 days at the equator (and slower near the poles) — so any midriff bulge isn't noticeable. I wish I could say the same for myself.

The Sun's regions: Caught between the core and the corona

The *photosphere* ("sphere of light") is the visible surface of the Sun (see Figure 10-2). When you glance at the bright disk of the Sun in the sky (be careful not to stare at it), you see the photosphere. When you see sunspots in a photograph of the Sun (or with a telescope as I describe later in this chapter), you're looking at a photo or a live view of the photosphere. And when I write that the diameter of the Sun is 864,500 miles, as I do in the previous section, I refer to the size of the photosphere. The temperature of the photosphere is 9,900°F (5,500°C).

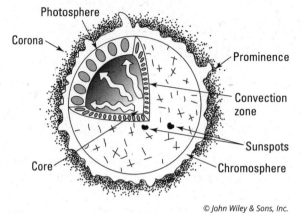

FIGURE 10-2:
The Sun is a hotbed of activity as it powers its piece of the universe.

Labels: Photosphere, Corona, Prominence, Convection zone, Sunspots, Chromosphere, Core

© John Wiley & Sons, Inc.

Above the photosphere are the two main outer layers of the Sun:

TIP

» **Chromosphere ("sphere of color"):** A thin layer that you can see during a total eclipse of the Sun, when it becomes visible as a narrow red band around the dark limb of the Moon. (I describe solar eclipses in "Experiencing solar eclipses" later in this chapter.) The chromosphere measures only about 1,000 miles (1,600 kilometers) thick, but its temperature reaches 18,000°F (10,000°C).

You can view the chromosphere at the edge of the Sun if you use an expensive H-alpha filter, mentioned in the sidebar "Solar-viewing styles of the big spenders," or you can see it on images taken with professional telescopes and displayed on the NASA website (see the section "Looking at solar pictures on the Net") and on various professional observatory websites. And you may glimpse the chromosphere during a total eclipse of the Sun, covered later in this chapter.

The transition from the chromosphere to the hundred-times-hotter corona occurs in a very thin boundary layer called the *transition region*. It doesn't show up in views of the Sun.

» **Corona:** The largest and least dense (most tenuous) layer of the Sun. You see it as a pearly white region extending out from the eclipsed disk of the Sun during a total eclipse. The corona's shape is always different from day to day (as photographed by Sun-watching satellites) and from one total eclipse to the next (as seen by viewers on Earth). The corona doesn't have a particular diameter — it thins out gradually with distance from the photosphere, and the size you measure for the corona depends on the sensitivity of your measuring instrument. The more sensitive the instrument, the more of the corona you can detect. The corona is very thin and very hot — a sizzling 1.8 million degrees Fahrenheit (1 million degrees Celsius) and, in some places, even hotter.

The corona is so rarified and electrified that the Sun's magnetic field determines its shape. Where lines of magnetic force stretch and open outward into space, the coronal gas is thin and barely visible. It can readily escape in the form of solar wind (see the section "Solar wind: Playing with magnets"). Where lines of magnetic force reach up in the corona and then turn back down to the surface, they confine the coronal gas. The coronal region is thicker and brighter there. Some loop-shaped structures extend from the photosphere up into the corona and hold gas much cooler than the surroundings. These loops are called *prominences,* which you can see on the limb of the Sun during a total eclipse.

Solar interior is the collective name for everything below the photosphere, and it contains the following three main regions:

» **Core:** This region extends from the center of the Sun about 25 percent of the way to the photosphere (about 108,000 miles [174,000 kilometers]). Solar energy is generated in the core by nuclear fusion at high temperature and high density. Because temperature and density are greatest at the center and gradually decrease with distance outward, most solar energy comes from the innermost part of the core, and less from its outer parts. The energy is generated in the form of gamma rays (a type of light) and also as neutrinos, strange subatomic particles that I describe in the later section "Solar CSI: The mystery of the missing solar neutrinos." The gamma rays bounce off one atom to another, back and forth, but on average, they move upward and outward. The neutrinos zip right through the whole Sun and fly out into space. The farther out in the solar interior, the cooler the temperature gets.

» **Radiative zone:** This region extends from the outer boundary of the core to about 71 percent of the way from the center to the photosphere (307,000 miles [494,000 kilometers] from the center). This layer gets its name from the fact that much of the solar energy travels outward through the zone in the form of electromagnetic radiation (a physics term for light).

» **Convection zone:** This region begins at the top of the radiative zone, 307,000 miles (494,000 kilometers) from the center of the Sun, and ends just below the photosphere. Swirling currents of hot gas transfer energy in this zone, with hot currents rising from the bottom, cooling toward the top of the zone, and then falling down again, warming, and rising anew. (The same process brings heat from the bottom of a kettle of boiling water up to the surface.)

Solar activity: What's going on out there?

The term *solar activity* refers to all kinds of disturbances that take place on the Sun from moment to moment and from one day to the next. All forms of solar activity, including the 11–year sunspot cycle and some even longer cycles, seem to involve magnetism. Deep inside the Sun, a natural dynamo generates new magnetic fields all the time. The magnetic fields rise to the surface and to higher layers in the solar atmosphere. They twist around and cause sunspots, eruptions, and other changing phenomena.

Astronomers measure magnetic fields on the Sun by their effects on solar radiation, using instruments called *magnetographs*. You can see images taken with these devices on some professional solar observatory websites (see the section "Looking at solar pictures on the Net"). These magnetic field observations show that sunspots are areas of concentrated magnetic fields and that sunspot groups have north and south magnetic poles. Outside of sunspots, the overall magnetic field of the Sun is pretty weak.

Many of the rapidly changing features on the Sun and probably all explosions and eruptions seem to be related to solar magnetism. Where there are changing magnetic fields, electrical currents occur (as in a generator), and when two magnetic fields bump into each other, a short circuit — called a *magnetic reconnection* — can suddenly release huge amounts of energy.

I cover several types of solar activity in the following sections.

Coronal mass ejections: The mother of solar flares

For decades, astronomers believed that the main explosions on the Sun were *solar flares.* They thought that solar flares occurred in the chromosphere (covered earlier in this chapter) and set things off.

Now astronomers know that they were just like the blind man who feels an elephant's tail and thinks that he knows all about the beast when he's actually touching one of the animal's less significant parts. Observations from space reveal that the primary engines of solar outbursts aren't solar flares, but *coronal mass ejections* — huge eruptions that occur high in the corona. Often a coronal mass ejection triggers a solar flare beneath it in the low corona and chromosphere. You can see solar flares in many of the images on professional astronomy websites. As the number of sunspots increases over an 11-year sunspot cycle (see the following section), so does the number of flares.

Scientists didn't know about coronal mass ejections for many years because they couldn't see them. Astronomers could get a good view of the corona only at rare intervals during the brief duration of a total eclipse of the Sun (see the section "Experiencing solar eclipses" later in this chapter). But solar flares can be seen at any time, so scientists studied them intensely and overestimated their importance.

You can observe prominences (see the previous section) on the edge of the Sun even when there's no total eclipse, but you need an expensive H-alpha filter (I describe these filters in the sidebar "Solar-viewing styles of the big spenders.") If you make enough observations, you will see that prominences sometimes erupt.

These eruptive prominences may also be stages in the development of coronal mass ejections.

When satellite images show a coronal mass ejection that isn't going off, say, to the east or to the west from the Sun, but that forms a huge expanding ring or *halo event* around the Sun, that's bad news. The halo event signifies that the coronal mass ejection — about a billion tons of hot, electrified, and magnetized gas — is heading right at Earth at about a million miles per hour. When it strikes Earth's magnetosphere (which I describe in Chapter 5), dramatic effects sometimes result, as I describe later in the section "Solar wind: Playing with magnets."

TIP

If you see a halo event in one of the satellite images, check the National Oceanographic and Atmospheric Administration (NOAA) Space Weather Prediction Center website (www.swpc.noaa.gov); NOAA may be forecasting some pretty fierce space weather. And the same website may display the latest video of a coronal mass ejection photographed from a satellite.

Cycles within cycles: The Sun and its spots

Sunspots are regions in the photosphere where the magnetic field is strong. They appear as dark spots on the solar disk (see Figure 10-3). The spots are cooler than the surrounding atmosphere and often appear in groups.

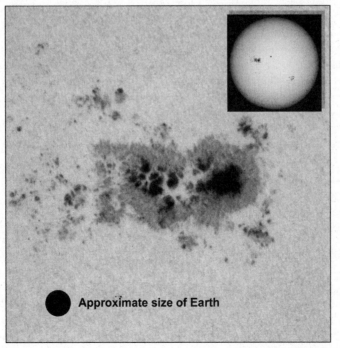

FIGURE 10-3:
A sunspot group 12 times bigger than Earth was photographed on September 23, 2000.

Approximate size of Earth

Courtesy of SOHO, NASA/ESA

The number of sunspots on the Sun varies dramatically over a repeating cycle that lasts about 11 years — the famous *sunspot cycle.* In the past, people blamed everything from bad weather to a decline in the stock market on sunspots. Usually, 11 years pass between successive peaks (when the most spots occur) of the sunspot cycle, but this period can vary. Furthermore, the number of spots at the peak may differ widely from one cycle to the next. Experts are always predicting how many sunspots will appear in the next cycle, but these long-range forecasts aren't very reliable.

As a sunspot group moves across the solar disk due to the Sun's rotation, the biggest spot on the forward side (the part of the group that leads the way across the disk) is called the *leading spot.* The biggest spot on the opposite end of the group is the *following spot.*

Magnetograph observations show definite patterns in most sunspot groups. During one 11-year cycle, all the leading spots in the Northern Hemisphere of the Sun have north magnetic polarity, and the following spots have south magnetic polarity. At the same time in the Southern Hemisphere, the leading spots have south polarity, and the following spots have north polarity.

Here's how these polarities are defined: The compass needle that points north on Earth is called a *north-seeking compass.* A *north magnetic polarity* on the Sun is one that a north-seeking compass would point to, if there was a compass on the Sun and it didn't melt. In the same sense, a *south magnetic polarity* on the Sun is one that a north-seeking compass would point away from.

Just when you think that you have it straight, guess what? A new 11-year cycle begins, and the polarities all reverse. In the Northern Hemisphere, the leading spots have south polarity, and the following spots have north polarity. In the Southern Hemisphere, the magnetic polarities reverse, too. If you were a compass, you wouldn't know whether you were coming or going.

To encompass all this information, astronomers have defined the *Sun's magnetic cycle.* The cycle is about 22 years long and contains two sunspot cycles. Every 22 years, more or less, the whole pattern of changing magnetic fields on the Sun repeats itself.

The solar "constant": Time to face the changes

The total amount of energy the Sun produces is called the *solar luminosity.* Of greater interest to astronomers is the amount of solar energy that Earth receives, or the *solar constant.* Defined as the amount of energy falling per second on 1 square meter of area facing the Sun at the average distance of Earth, the solar constant amounts to 1,368 watts per square meter (127 watts per square foot).

Measurements made by solar and weather satellites that NASA sent up in the 1980s revealed very small changes in the solar constant as the Sun turns. You may think that Earth receives less energy when dark sunspots are present on the solar disk, but that isn't the case. In fact, the opposite is true: more sunspots, more energy Earth receives from the Sun. Chalk up another mystery for astronomers to solve.

According to astrophysical theory, the Sun was somewhat brighter when it was very young than it has been for the last several billion years. Theory also predicts that the Sun will cast much more energy on Earth ages from now, when it becomes a red giant star (see Chapter 11).

So "solar constant" sounds like wishful thinking, although from day to day, any change in the energy output of the Sun is extremely small.

Solar wind: Playing with magnets

An electrified gas or "plasma" called the *solar wind* is always flowing away from the solar corona. It moves through the solar system at about a million miles per hour (470 kilometers per second) as it passes Earth's orbit.

The solar wind comes in streams, fits, and puffs and constantly disturbs and replenishes Earth's magnetosphere, which becomes compressed in size and swells again. The disturbances to the magnetosphere, especially those from traveling solar storms such as coronal mass ejections (covered earlier in this chapter), can cause displays of the Northern Lights (aurora borealis) and Southern Lights (aurora australis), as well as geomagnetic storms (see Chapter 5 for more on the magnetosphere and auroras). The geomagnetic storms can shut down power company utility grids (causing blackouts), blow out electronic circuits on oil and gas pipelines, interfere with GPS and radio communications, and damage expensive satellites. Some people even claim they can *hear* aurorae, and a 2011 experiment in Finland recorded these sounds for the first time.

Coronal mass ejections (CMEs) are usually invisible with amateur equipment but marvelously revealed by satellite telescopes. They are billion-ton blobs of solar plasma, permeated with magnetic fields, that sweep out into the solar system. Sometimes CMEs collide with Earth's magnetosphere — the huge region where electrons, protons, and other electrically charged particles bounce back and forth from high northern latitudes to high southern latitudes, trapped in Earth's magnetic field. The magnetosphere acts as a protective umbrella against coronal mass ejections and the solar wind.

But the protection is less than perfect. Sometimes CMEs or other space weather damages expensive satellites in orbit around Earth. In other cases, they give an added radiation dose to people flying on polar routes, interfere with radio communications, or even harm the electrical grid or electronic equipment on pipelines. Society is becoming more aware of the hazards of space weather. In 2016, the *Wall Street Journal* reported that insurers are offering "catastrophe bonds" to cover some of their clients' potential losses from natural disasters, including a solar flare.

TIP

Solar disturbances and their effects on the magnetosphere are called *space weather.* You can see the latest official U.S. government space weather report and forecast at the website of the NOAA Space Weather Prediction Center (www.swpc.noaa. gov). If you're willing to invest about $2, the Space Weather App for iPhone and Android gives you space weather forecasts and alerts for auroras, radio blackouts, and other possible effects of space weather.

Solar CSI: The mystery of the missing solar neutrinos

The nuclear fusion at the heart of the Sun does more than change hydrogen into helium and release energy in gamma rays to heat the whole star. It also releases enormous numbers of *neutrinos*, or electrically neutral subatomic particles that have almost no mass, travel at nearly the speed of light, and can pass through almost anything. Astronomers can check on the calculated temperature and density in the core of the Sun by observing neutrinos that come from the star.

A neutrino is like a hot knife through butter: It easily cuts through. In fact, neutrinos can fly right out from the center of the Sun and into space. Those that head Earthward fly right through Earth and out the other side. But the neutrino is different from the hot knife because the knife also melts butter that it comes in contact with. The neutrino just whooshes through without affecting the matter it passes through in almost (but not quite) every case.

Physics experiments can detect the rare exceptions in which neutrinos do interact with matter, so a tiny fraction of the solar neutrinos that pass through huge underground laboratories known as neutrino observatories do get counted. These observatories are located mostly in deep mines and tunnels under mountains. At such depths, few other kinds of particles fly around, so scientists have an easier time telling a solar neutrino from other particles. One major facility, the Sudbury Neutrino Observatory in Canada, is 6,800 feet below Earth's surface. It's a good place to "delve deeply" into astronomy.

Counting neutrinos isn't easy, but some time ago, reports from the neutrino observatories indicated a deficiency in solar neutrinos: The number of neutrinos coming to Earth was significantly fewer than the number scientists expected, based on the rate at which the Sun generates energy.

The solar neutrino deficiency was the least of our problems on Earth. It paled in significance beside AIDS, war, famine, the depletion of the forests, the extinction of valuable species, and the consumption of irreplaceable fossil fuel reserves. But the loss nagged at scientists, prompting them to make new theories of particle physics and to check on theoretical models of the solar interior.

Fortunately, scientists at the Sudbury Neutrino Observatory (and elsewhere) solved the problem of the missing neutrinos. It turns out that some of the neutrinos produced in the Sun's core change to either of two other types of neutrinos on the way to Earth, and earlier neutrino observatories, which reported the solar neutrino deficiency, couldn't detect the two other types. The problem was that our ignorance of how neutrinos behave led to deficiencies in the design and capabilities of the laboratory equipment. It was not, after all, a misunderstanding of how the Sun generates energy or how many neutrinos it emits. Here's a good analogy: Suppose you count birds for a wildlife survey, but you wear eyeglasses with colored lenses. The colored glasses make it difficult to see birds of certain colors, so you may think bluebirds are endangered, but the problem is that you can see only cardinals.

Four billion and counting: The life expectancy of the Sun

Someday the Sun must run out of fuel, so someday it will die. And without the energy and warmth of the Sun, life on Earth would cease to exist. The oceans would freeze, and so would the air. Seems logical, right? But what will *actually* happen is that the Sun will swell up and take the form of a red giant star (see Chapter 11 for more about red giants). It will look enormous, and it will fry the oceans. So the oceans will actually evaporate before they have a chance to freeze.

Read the preceding paragraph carefully: I didn't say that the oceans *will* freeze; I said that they *would* freeze without the energy of the Sun. In fact, the energy Earth receives will increase so much before the Sun dies out that humans will die of the heat (if people still exist), not of the cold. And as for the seas, boiled tuna will be served, not frozen cod. Talk about global warming!

The red giant Sun will puff off its outer layers, forming a colorful, expanding gas cloud that astronomers call a planetary nebula. But no humans will be left to admire it. To appreciate what we'll surely miss, take a good look at planetary nebulas created by other stars (see Chapter 12).

The nebula will gradually fade away, and all that will remain at its center is a tiny cinder of the Sun, a hot little object called a *white dwarf star.* It won't be much larger than Earth, and although it will be very hot at first, it will be too small to cast much energy on Earth. Whatever's left on the surface of Earth will freeze. And the white dwarf will cool down and fade out like an ember in a dying campfire. (For more on white dwarfs, flip to Chapter 11.)

Fortunately, we should have about 5 billion years to go before that prospect looms near. Future generations can worry about this, along with the national debt and how to acquire rare first editions of *Astronomy For Dummies.*

Don't Make a Blinding Mistake: Safe Techniques for Solar Viewing

Seventeenth-century Italian astronomer Galileo Galilei made the first great telescopic discovery about the Sun. By watching the daily movements of sunspots across the solar surface, he deduced that the Sun rotates. By some accounts, he damaged his eyesight while researching the Sun. Those stories may be wrong, but my warning isn't: Looking at the Sun through a telescope or another optical aid such as binoculars is extremely dangerous. A telescope or a pair of binoculars collects more light than the naked eye and focuses it on a small spot on your retina, where it will cause immediate, severe harm. Ever see a *burning glass,* a magnifying lens that focuses the rays of the Sun on a piece of paper to set it on fire? Now you get the idea.

WARNING

Looking at the Sun with the naked eye isn't a good idea, and in some cases, it can be harmful. Taking even the briefest peek at the Sun through a telescope, binoculars, or any other optical instrument (whether you use your property or someone else's) is very dangerous unless the device is equipped with a properly installed solar filter made by a reputable manufacturer specifically for viewing the Sun. However, you can observe the Sun with a technique called *projection* (see the following section). If you carefully follow the instructions in the next two sections, you likely won't have a bad experience. Better yet, start your solar observing under the guidance of an experienced amateur or professional astronomer. (Head to Chapter 2 to find out about clubs and other resources that help you get going.)

Viewing the Sun by projection

Galileo invented the *projection technique* by using a simple telescope to cast an image of the Sun on a screen in the manner of a slide projector. This technique is safe only when used properly with simple telescopes, such as those sold under the description *Newtonian reflector* or *refractor.*

As I explain in Chapter 3, a Newtonian reflector uses only mirrors, aside from the eyepiece, and its eyepiece is near the top of the telescope tube, protruding at right angles. A refractor works with lenses and doesn't contain a mirror.

Don't use the projection technique with telescopes that incorporate lenses and mirrors along with eyepieces. In other words, don't use the projection technique with the Schmidt-Cassegrain or Maksutov-Cassegrain telescope models, which use both mirrors and lenses (I describe these telescopes in Chapter 3). The hot, focused solar image may damage the apparatus inside the sealed telescope tube and can pose a danger.

When you get a good handle on the projection technique, you can look for sunspots. If you spot some spots, look again tomorrow and the next day to monitor their movement across the solar disk. In reality, although they may move a bit on their own, most of the sunspots' movements are due to the turning of the Sun, or the *solar rotation*. You're repeating Galileo's discovery in the safe manner that he pioneered.

Using the projection technique with a Newtonian telescope

Here's how to safely view the Sun with the projection technique:

1. **Mount a Newtonian reflector or a refractor telescope on a tripod.**

2. **Install your lowest-power eyepiece in the telescope.**

3. **Point the telescope in the general direction of the Sun *without* sighting through or along the telescope; keep yourself and all other people away from the eyepiece and not behind it where the focused solar beam emerges.**

   If the telescope has a small finder telescope mounted on it, don't look through the finder telescope either!

4. **Find the shadow of the telescope tube on the ground.**

5. **Move the telescope up and down and back and forth while watching the shadow, to make it as small as possible.**

   The best way is for you or an assistant to hold a piece of cardboard beneath the telescope, perpendicular to the long dimension of the scope, so the shadow of the tube falls on the cardboard. Move the telescope so the tube shadow is as close to a solid, dark, circular shape as possible.

6. **Hold the cardboard at the eyepiece; the Sun will be in the field of view, and its image will project onto the cardboard.**

If the Sun's image isn't in view, the bright glare of the Sun should be visible on one side of the cardboard; in that case, move the telescope to move the glare toward the center of the cardboard, thus moving the Sun into view. Keep in mind that moving the cardboard farther away from the eyepiece will make the projected image of the Sun bigger and easier to view. But if you move the cardboard too far out, the image will be too dim for viewing.

TIP

Figure 10-4 presents a diagram of the projection technique. (*Note:* The diagram shows a small finder scope mounted on the telescope because most telescopes are equipped that way, but you *must not* look at the Sun through the finder scope or the main telescope because severe eye damage will occur.) The easiest and safest way to practice the technique is to consult an experienced observer from your local astronomy club; flip to Chapter 2 to find out how to locate a club in your area.

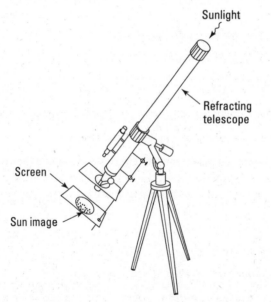

Sunlight

Refracting
telescope

Screen

Sun image

FIGURE 10-4:
Project the Sun
onto a white
surface to protect
your eyes.

Dinah L. Moché/Astronomy: A Self-Teaching Guide, Seventh Edition

Even though you avoid looking through the telescope, you have to beware of other hazards the projection method presents. I once saw a student in Brooklyn project a solar image with a 7-inch telescope. He knew better than to look through the eyepiece. But he carelessly moved his arm across the projected sunlight close to the eyepiece, where the solar image is very small. The hot concentrated image burned a smoking hole in his black leather jacket.

You must take great care when using a telescope as a solar image projector, and you must *never* allow unsupervised children or any person who isn't trained in this method to operate the telescope. Don't look at the Sun through your telescope, and don't look at the Sun through the small finder telescope or viewfinder that your telescope may be equipped with. To avoid injury, make sure that no part of anyone's body, clothing, or other property gets in the projected beam of sunlight; only your cardboard projection screen should be in the beam.

Science teachers who want to show sunspots to their classes should consider perhaps the safest option: a telescope that is designed to serve only one purpose, projecting a white light image of the Sun. An instrument that seems to fill the bill is the Sunspotter (about $370), available at www.teachersource.com.

If you don't want to use the projection technique, or if you have the kind of telescope that uses both lenses and mirrors — which you shouldn't use with this technique — you can still view the Sun safely, but you need a special *white light* solar filter. "White light" means the filter admits all or most of the colors of visible light; "filter" means that it cuts down on the brightness of the light. Viewing the Sun through a safe white light filter (see the "Viewing the Sun through front-end filters" section) requires a modest investment, but the price you pay is well worth the viewing and the safety.

Making a pinhole camera to project the Sun

You can use a pinhole camera as an inexpensive way to safely view the Sun (but not in as much detail as the methods described in the preceding section) that's said to have been used since the Dark Ages, long before Galileo and the telescope came around. NASA's Jet Propulsion Laboratory tells you how to do it with two pieces of white cardboard, a piece of aluminum foil, tape, and a pin or a paper clip. See the illustrated step-by-step directions at www.jpl.nasa.gov/edu/learn/project/how-to-make-a-pinhole-camera/.

Believe it or not, you'll project an image of the Sun with just the hole made by a pin, no lens or mirror required! This technique is fun for viewing a partial eclipse of the Sun or the partial phases of a total eclipse, and it will also show large sunspot groups when they appear on the solar disk, but little else. It's a good project for a family, elementary school class, or scout troop.

After you've made your first pinhole camera, you can experiment with building better ones, using a cardboard box or a good-sized mailing tube to keep stray light out and give the image better contrast.

Viewing the Sun through front-end filters

The only solar filters that I recommend for observing the Sun in ordinary white light go at the *front end* of your telescope so no light can enter the telescope without passing through the filter. (These filters aren't the H–alpha filters that I describe in the sidebar "Solar-viewing styles of the big spenders," and they're much less expensive.) White light filters show sunspots, which H–alpha filters don't. But white light filters don't show solar prominences and flares, which you can see through H–alpha filters.

Filters at, near, or in place of the eyepiece may break as a result of concentrated solar heat, causing possibly great damage to your eye. Use only filters that go at the front end of your telescope.

I give you the scoop on various types of scopes in Chapter 3, and I recommend the following front-end telescope filters for solar-viewing use:

>> **Full-aperture filters:** Appropriate for telescopes of 4-inch aperture or less (the *aperture* is the diameter of the light-collecting mirror or lens in your telescope), such as the Meade ETX-90 and the Celestron SkyProdigy 90 (see Chapter 3). The filter extends across the full diameter of the telescope so the entire light-collecting mirror or lens receives the filtered light from the Sun.

>> **Off-axis filters:** Best for telescopes of 4-inch aperture or more — but not for refractors. An off-axis filter is smaller than the aperture of the telescope, but it's mounted in a plate that covers the entire aperture. The Sun is so bright that you don't need the whole aperture of the telescope to collect enough light for good solar viewing. A larger aperture potentially can give a sharper view, but in most viewing locations, blurring by Earth's atmosphere nullifies this advantage. The less unneeded sunlight gets into your telescope, the safer you and the telescope will be.

You want an off-axis solar filter with most telescopes other than refractors because nonrefractors usually have little mirrors or mechanical devices on-center inside the telescope tube, blocking the part of the light that comes down the center of the tube.

In the special case of a refractor of 4-inch aperture or more, your filter needs to go over the top end of the telescope and be smaller than the telescope aperture, but it should mount centrally in an opaque plate that covers the telescope. The filter should mount on-center because, generally, the central part of the primary or objective lens of the refractor (the big lens) may have better optical quality than the periphery of the lens.

Thousand Oaks Optical in California manufactures full-aperture and off-axis glass solar filters of several kinds. The company sells filters made to fit many different specific commercially available telescopes, as listed on its website at www.thousandoaksoptical.com/solar.html.

Use solar filters only in accordance with the manufacturer's directions.

Fun with the Sun: Solar Observation

The Sun is a fascinating, constantly changing ball of hot gases that offers plenty of viewing opportunities for the prudent observer. In addition to observing the Sun yourself (using the precautions described earlier in the chapter), you can visit websites that offer awe-inspiring, professionally produced pictures. And taking advantage of both provides you with the full solar-viewing experience. This section suggests some ways that you can personally enjoy old Sol.

Tracking sunspots

When you become confident in your ability to observe the Sun safely with the projection method or by equipping your telescope with a safe solar filter, you can begin to study sunspots using the following plan:

>> Observe the Sun as often as possible.

>> Note the sizes and positions of sunspots and groups of sunspots on the solar disk.

Some sunspots look just like tiny dark spots. If the spot is truly a tiny dark spot, even through a powerful observatory telescope, it's a *pore*. But if a sunspot is big enough, you can distinguish its different regions. The dark central portion is called the *umbra,* and the surrounding area that appears darker than the solar disk but lighter than the umbra is the *penumbra.*

>> Chart the motion of the sunspots as the Sun makes one complete turn — which takes 25 days (at the equator) to about 35 days (near the poles; yes, the Sun turns at different rates at different latitudes, another one of its many mysterious and unexpected properties).

The Solar Section of the Association of Lunar and Planetary Observers offers an excellent free form for recording and reporting sunspots at alpo-astronomy. org/solarblog/?page_id=920. The British Astronomical Association's Solar Section also offers solar observing forms. Visit www.britastro.org/solar/ index.php?style=orig.

As you track sunspots, you may want to note how many you see in a day; that figure is called (guess what?) the *sunspot number.* You may even want to keep track of the sunspot number from year to year to see if you can measure the sunspot cycle yourself. In the following sections, I give you info on how to compute the sunspot number and where to find official numbers.

Figuring your personal sunspot number

Compute your own sunspot number for each day of observation, using this formula:

$$R = 10g + S$$

R is your personal sunspot number, g is the number of groups of sunspots you see on the Sun, and s is the total number of sunspots you count, including the spots in groups. Sunspots usually appear isolated from each other on different parts of the solar disk. Spots close together on one part of the disk are a group. A completely isolated spot counts as its own group (the reasoning behind this designation is pretty spotty, but that's the way scientists have done it for years).

Suppose you count five sunspots; three are close together in one place on the Sun, and the other two appear at widely separated locations. You have three groups (the group of three and the two groups consisting each of one spot), so g is 3. The number of individual spots is five, so s is 5.

$$R = (10 \times 3) + 5$$
$$R = 30 + 5$$
$$R = 35$$

Finding official sunspot numbers

On the same day, different observers come up with different personal sunspot numbers. If you have better viewing conditions and a better telescope, or maybe just a better imagination, you calculate a higher sunspot number than your neighbor Jones. You calculate $R = 59$, and that bum Jones can only claim $R = 35$. When it comes to sunspot numbers, you're way ahead! Now when it comes to whose lawn is nicer, well, I'll leave that debate up to your neighbors. "R" you clear on this?

Central authorities who tabulate and average the reports from many different observatories find by experience that some observers keep pace with Jones, some can't see as many, and some, like you, are far ahead. From this experience, the authorities calibrate each observatory or observer and make allowances in future counts so they can average the reports and get the best estimate of the sunspot number for each day.

You can check out the sunspot number every day (or whenever you want) at www.spaceweather.com.

Experiencing solar eclipses

On a daily basis, the best way to see the Sun's outermost and most beautiful region, the corona, is to view the satellite images posted on the websites I list in the next section. But seeing the corona "live and in person" is a spectacle that you shouldn't deny yourself. The corona during a total eclipse of the Sun is one of nature's finest spectacles. It's why many amateur astronomers save their earnings for years to splurge on a great eclipse trip (see Chapter 2 for details). Professional astronomers make their way to the eclipse, too, even though they have satellites and space telescopes at their disposal.

The Sun experiences *partial, annular,* and *total eclipses*. The greatest spectacle is the total eclipse; some annular eclipses are well worth the trip, too. (During an annular eclipse, a thin, bright ring of the photosphere is visible around the edge of the Moon.) A partial eclipse isn't something to drive hundreds of miles out of your way to see because you don't see the chromosphere or corona, but you definitely want to check it out if one comes your way. After all, the first and last stages of a total eclipse or an annular eclipse are partial eclipses, so you need to know how to observe those stages, too.

Observing an eclipse safely

To observe a partial eclipse, or the partial eclipse phases of a total eclipse, use the solar filters I describe in the section "Viewing the Sun through front-end filters" earlier in this chapter. You can watch through binoculars or telescopes equipped with filters, you can hold a filter in front of your eyes, or you can use the technique I describe in the earlier section "Viewing the Sun by projection."

A total eclipse normally starts with a partial phase, beginning with *first contact,* when the edge of the Moon first comes across the edge of the Sun. You now see a *partial eclipse* of the Sun, signifying that you're in the *penumbra,* or light outer shadow, of the Moon (see Figure 10-5).

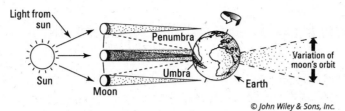

© John Wiley & Sons, Inc.

At *second contact*, the Moon's leading edge reaches the far edge of the Sun, totally blocking the Sun from your view. Now you witness a *total eclipse*; you're in the dark *umbra*, or central shadow, of the Moon. You can put down your viewing filter or your filtered binoculars and stare safely at the fantastic sight of the totally eclipsed Sun. But you must not stare at the Sun when totality is over — it's unsafe to do so.

The corona forms a bright white halo around the Moon, perhaps with long streamers extending east and west. You may see thin bright polar rays in the corona off the north and south limbs of the Moon. Watch for small, bright red points, which are solar prominences that sometimes are visible to the naked eye during brief moments of the eclipse. A thin red band along the limb of the Moon is the solar chromosphere, better seen at some total eclipses than at others. Near the peak of the 11-year sunspot cycle, the corona is often round, but near the sunspot minimum part of the cycle, the corona is elongated east to west. The corona takes a different shape during every eclipse.

WARNING

Some people take the solar filters off their binoculars or telescopes and look at the totally eclipsed Sun through these instruments without the benefit of filters. This approach is very dangerous if

>> You look too soon, before the Sun is totally eclipsed.

>> You look too long (a very easy way to have an accident) and continue to watch through an optical instrument after the Sun begins to emerge from behind the Moon.

Be careful! I strongly advise against telescopic and binocular viewing of the Sun without filters, even during total eclipse, unless you're viewing under the direct control of an expert. Sometimes, for example, the experienced leader of an eclipse trip or cruise group uses a public address system, computer calculations, and personal observing know-how to announce when you can look at the eclipsed Sun. The leader also tells you when you must stop, with plenty of warning.

NOT YOUR DADDY'S 3-D SHADES

Rainbow Symphony sells several types of Eclipse Glasses that resemble the 3-D viewers you wear like eyeglasses in movie theaters. They're intended for viewing partial phases of a total eclipse or any phase of a partial solar eclipse. Check them out on the Rainbow Symphony website at www.rainbowsymphony.com. Thousand Oaks Optical (www.thousandoaksoptical.com) offers a similar product that it calls Solar Viewers. These manufacturers' products are relatively inexpensive, but the catch is that you may have to buy them in bulk. If you take family or friends along to a solar eclipse, I suggest bringing enough frames for everyone.

In my experience (which was painful), the easiest way to hurt yourself is to linger looking through your binoculars or telescope for "just another second" when a tiny part of the bright visible surface of the Sun starts to emerge from behind the Moon. That tiny bright part may not make you immediately avert your vision because it doesn't seem brilliant enough. But what you don't realize is that the infrared rays (which are invisible) from the small, exposed part of the solar surface damage your eyes without dazzling them or causing immediate pain. In a few minutes or less, you begin to feel the pain. By then, the damage is done.

If you observe safely, follow all directions, and never take chances looking at the Sun, you can look forward to many happy returns of total eclipses of the sun!

Seeking shadow bands, Baily's Beads, and the diamond ring

Another good reason to avoid looking at the Sun with optical instruments during the total phase of an eclipse is that you have so much to look for all around the sky with your naked eye.

Here are some neat phenomena you can look for during an eclipse:

>> Just before totality, so-called *shadow bands,* shimmering, low-contrast patterns of dark and light stripes, may race across the ground or, if you're at sea, across the deck of your ship. The stripes are an optical effect produced in Earth's atmosphere when the bright disk of the Sun dwindles to the last little sliver behind the eclipsing Moon before becoming completely eclipsed.

>> *Baily's Beads* occur just instants before and after totality, when little regions of the bright solar surface shine through gaps between mountains and crater rims on the edge of the Moon. At one moment, there may be just one very bright bead. Astronomers call this aspect of the eclipse the *diamond ring.*

(The bright inner corona looks like a narrow ring around the Moon, and the bright bead is the "diamond.")

>> Wild animals (and domesticated animals, if you're near a farm) react notably to the eclipse. Birds come down to roost, cows head back for the barn, and so on. During one 19th-century eclipse, some top scientists set up their instruments in a barn, pointing the telescopes out through the door. Boy were they surprised when totality began and the livestock ran in!

>> Where sunlight shines down through the leaves of a tree, you often see a dappled pattern on the ground below — small bright spots, shaped like the Sun. Before the eclipse begins, the dapples are round, like the uneclipsed Sun. During the partial eclipse, the dapples look like half Moons and then crescents, changing in shape according to the current phase of the eclipse. You can observe dapples like this without a tree; just bring a kitchen colander and let the Sun shine through the many little holes during an eclipse. Or if you made a pinhole camera (see the earlier "Making a pinhole camera to project the Sun" section), you can use it to view the eclipse.

TIP

When the Sun becomes totally eclipsed, look at the dark sky all around the Sun. You have a rare chance to see stars in the daytime. Special articles published in astronomy magazines or posted on their websites tell you which stars and planets to look for at each total solar eclipse. Or you can figure it out yourself by simulating the sky at the date and time of the eclipse on your desktop planetarium program or similar smartphone app (see Chapter 2). All you need to do is set the program to display the sky as it will be from the place where you expect to observe.

Following the path of totality

Totality ends at *third contact*, when the Moon's trailing edge moves out across the solar disk. At the last moment of totality, a small bright area of the photosphere may emerge from behind the Moon. Then you see the diamond ring that I describe in the preceding section. Now you're back in the penumbra, and you can see a partial eclipse. At *fourth* or *last contact*, the Moon's trailing edge moves off the forward limb of the Sun. The eclipse is over.

The whole eclipse, from first contact to last contact, may take a few hours, but the good part, totality, lasts from less than a minute to seven minutes or slightly more.

One place on the *path of totality* — the track of the center of the Moon's shadow across the surface of Earth — boasts the greatest duration. Totality is briefer everywhere else on the path. Of course, the place where the eclipse has maximum duration may not be the place where the weather prospects are best, or it may not be a place you can easily or safely reach. So advance planning of your eclipse trip is vital. At any good location, all the accommodations, rental vehicles, and so on are usually booked up at least a year or two in advance of the eclipse.

To plan your eclipse trip, pick a likely eclipse from Table 10-1 and start investigating the best way to view it.

TABLE 10-1 **Future Total Eclipses of the Sun**

Date of Total Eclipse	Maximum Duration (Minutes and Seconds)	Path of Totality
July 2, 2019	4:33	Across the South Pacific, Chile, and Argentina
December 14, 2020	2:10	Across the South Pacific, Chile, and Argentina, and across the South Atlantic
December 4, 2021	1:54	Across the Southern Ocean, Antarctica, the Weddell Sea, the Southern Ocean again, and to the South Atlantic
April 8, 2024	4:28	From the Pacific Ocean across Mexico and across the United States from Texas through the Midwest; New York; northern New England; New Brunswick, Canada; and out over the North Atlantic
August 12, 2026	2:18	From the Arctic Ocean to eastern Greenland, across the Atlantic Ocean to western Iceland, and northern Spain, and across the Mediterranean Sea
August 2, 2027	6:23	From the Atlantic Ocean across northern Africa to Egypt, Saudi Arabia, Yemen, and northwest Somalia, and across the Arabian Sea and Indian Ocean
July 22, 2028	5:10	From the Indian Ocean across Australia, over the Tasman Sea, and across the South Island of New Zealand into the South Pacific Ocean

TIP

A few years before each eclipse, articles with information on the weather prospects and logistics for viewing from various locations begin to appear in astronomy magazines. Check the *Sky & Telescope* and *Astronomy* magazine websites (see Chapter 2). Look for eclipse tour advertisements in the magazines and on the web. Check out the authoritative eclipse predictions on the NASA eclipse site, at `eclipse.gsfc.nasa.gov/solar.html`. See Chapter 2 for my advice on eclipse trips. And have a great time!

Looking at solar pictures on the Net

You can see current or recent professional photographs of the solar disk and sunspots (what solar astronomers call white-light photographs — white light being all the visible light of the Sun) at various places on the web. A good place to look is the site of Italy's Catania Astrophysical Observatory, `www.oact.inaf.it/weboac/sun/`. The white-light photo is the one labeled "Continuum," a technical term that means a colored filter wasn't used to take the picture. You can get experience in identifying a sunspot group and counting sunspots by practicing on these photos.

Sometimes the weather is cloudy in Catania, so you look elsewhere for a current professional white-light photo of the solar disk. One good place is the Full Disk Observations page of the Big Bear Solar Observatory in California, at `www.bbso.njit.edu/cgi-bin/LatestImages`. You can't see the corona yourself when there's no total eclipse, but you can see white-light images of the corona from SOHO, a satellite developed by the European Space Agency and NASA, at The Very Latest SOHO Images website (`soho.nascom.nasa.gov/data/realtime-images.html`). They're the images labeled "LASCO."

You can see recent maps of the solar magnetic field at the websites of some professional observatories, including Stanford University's Wilcox Solar Observatory (`wso.stanford.edu`). Another good source is the National Solar Observatory, on the web page at `solis.nso.edu/0/vsm/vsm_fulldisk.html`.

When you become an advanced amateur astronomer and want to photograph celestial scenes through your telescope, you may want to try solar photography. You can find inspiring examples at the Mount Wilson Observatory, where researchers have been photographing the Sun since 1905. Check out the fantastic picture of an airplane silhouetted against a spotty Sun and the picture of the largest sunspot group ever photographed, from April 7, 1947. If you're lucky enough to see a sunspot group even half as large, you may be able to see it not only with a telescope but also by looking through a solar filter (like those I describe in the nearby sidebar "Not your daddy's 3-D shades") without any other optical aid. The Mount Wilson site for a few historic white-light solar photographs is `physics.usc.edu/solar/direct.html`.

Astronomers study the Sun in all kinds of light, not just white light. Their research includes pictures taken in ultraviolet and extreme ultraviolet radiation and X-rays, which are all forms of light invisible to the eye and blocked by Earth's atmosphere. The pictures are made with telescopes mounted on satellites orbiting Earth at high altitude or taken by spacecraft located farther away and orbiting the Sun just as Earth does. Sun images from satellites and from many kinds of telescopes on the ground are available on NASA's Current Solar Images website, at `umbra.nascom.nasa.gov/newsite/images.html`.

Another great source of solar images is the Solar Dynamics Observatory (SDO), which was launched in 2010. Check out the spectacular images at the SDO website, sdo.gsfc.nasa.gov. And for videos of solar observations by the STEREO mission visit stereo.gsfc.nasa.gov.

If you're feeling like a solar astronomer now, try some of exercises on the Public Broadcasting System's *Sun* Lab website at www.pbs.org/wgbh/nova/labs/lab/sun/research. Just click on Solar Cycle and get started.

The Sun belongs to everyone, so study and observe it often. You'll be glad you did!

Chapter **11**

Taking a Trip to the Stars

The Sun is one of a few hundred billion stars in the Milky Way galaxy (also called "the Galaxy"), where Earth resides. Likewise, about two trillion other galaxies exist in the *observable universe*, meaning all of space as far as we can see at this point in time. Each of those galaxies contains a great many stars as well. And just like people, stars fit into dozens of classifications. But the overwhelming majority fall into several simple types. These types correspond to stages in the life cycles of stars, just as you classify people by their ages. (For more about the Milky Way and other galaxies, flip to Chapter 12.)

When you understand what a star is and how it runs through its life cycle, you get a feel for these shining beacons of the night sky — and the ones that aren't so bright, too.

In this chapter, I emphasize the initial mass of a star — the mass it's born with — as the main determinant of what the star will become. I continue with the key properties of stars, along with the features of binary, multiple, and variable stars that make them so interesting for you to observe.

And no discussion of stars is complete without some gossip about the celebrities, so I introduce you to some luminaries of the night sky that you'll want to know — the leading "personalities" of the solar neighborhood.

Life Cycles of the Hot and Massive

The most important star categories correspond to successive stages in a star's life cycle: baby, adult, senior, and the dying. (What! No teenagers? The universe gave up on youth classifications after the terrible twos.) Of course, no astrophysicist worth his PhD uses such simple terms, so astronomers refer to the stages of stars as young stellar objects (YSOs), main sequence stars, red giants, and those in the end states of stellar evolution, respectively. Many stars don't really die; they continue in a new state, becoming white dwarfs, neutron stars, or black holes. But in some cases, they're completely shattered.

Here's the life cycle of an ordinary star with about the same mass as the Sun:

1. **Gas and dust in a cool nebula condense, forming a young stellar object (YSO).**

2. **Shrinking, the YSO dispels its remaining birth cloud, and its hydrogen "fire" ignites.**

   In other words, nuclear fusion is underway, as I explain in Chapter 10.

3. **As the hydrogen burns steadily, the star joins the main sequence.**

   I describe this stage in stellar life in the section "Main sequence stars: Enjoying a long adulthood" later in this chapter.

4. **When the star uses up all the hydrogen in its core, the hydrogen in the shell (a larger region surrounding the core) ignites.**

5. **The energy released by the burning of the hydrogen shell makes the star brighter, and it expands, which makes its surface larger, cooler, and redder; the star has become a red giant.**

6. **Stellar winds blowing off the star gradually expel its outer layers, which form a planetary nebula around the remaining hot stellar core.**

7. **The nebula expands and dissipates into space, leaving just the hot core.**

8. **The core, now a white dwarf star, cools and fades forever.**

Stars with much higher masses than the Sun have different life cycles; instead of producing planetary nebulae and dying as white dwarfs, they explode as supernovas and leave behind neutron stars or black holes (or, in some cases, maybe nothing at all!). The life cycle of a massive star progresses rapidly; the Sun may last 10 billion years, but a star that begins with 20 or 30 times the Sun's mass explodes just a few million years after its birth.

Stars with much less mass than the Sun hardly have a life cycle. They begin as YSOs and join the main sequence to remain as red dwarfs forever. The explanation for this is a fundamental principle of stellar astrophysics: The bigger the mass, the fiercer and faster the nuclear fires burn; the smaller the mass, the less fiercely the fire burns, and the longer it lasts.

By the time our Sun uses up its core hydrogen, it will be at least 9 billion years old. But a red dwarf star burns hydrogen so slowly that it shines on the main sequence forever, for all practical purposes. (Given enough time, a red dwarf would exhaust its hydrogen fuel, but that amount of time is much greater than the present age of the universe, so all the red dwarfs that ever existed are still going strong now.)

The following sections describe the stellar stages in more detail.

Young stellar objects: Taking baby steps

Young stellar objects (YSOs) are newborn stars that are still surrounded or trailed by wisps of their birth clouds. The classification includes *T Tauri stars,* named for the first of their type — the star T in the constellation Taurus — and *Herbig-Haro objects,* named for the two astronomers who classified them. (Actually, H-H objects are glowing blobs of gas expelled in opposite directions from the young star, which is usually hidden from view by dust from its birth cloud.) YSOs form in stellar nurseries — called *HII regions* — such as the Orion Nebula (see Figure 11-1), where hundreds of stars have been born in the past 1 million or 2 million years.

FIGURE 11-1: The Orion Nebula cradles many young stellar objects.

Courtesy of ESO/Igor Chekalin

Often a YSO is at the center of a flattened cloud of gas and dust, called a *circumstellar disk*, which is feeding matter into the YSO, helping it form.

Many of the Hubble Space Telescope images of spectacular jetlike nebulae are pictures of YSOs. The jets and other nebular surroundings are prominent, but the stars themselves are sometimes barely visible (if you can see them at all), hidden by the surrounding gas and dust. (Flip to Chapter 12 for more about nebulae.) But some YSOs are much less prominent than the ones in the Hubble pictures, and astronomers need citizen scientists like you to help find them, as I discuss in the later section "Star Studies to Aid with Your Brain and Computer."

Main sequence stars: Enjoying a long adulthood

Main sequence stars, which include our Sun, have shed their birth clouds and now shine thanks to the nuclear fusion of hydrogen into helium that goes on in the core. A star with one solar mass takes about 50 million years to reach this state. More massive stars take much less time to reach the same point, and stars less massive than the Sun take much more time. (See Chapter 10 for more about nuclear fusion in the Sun.) For historical reasons, going back to when astronomers classified stars before they understood their differences, main sequence stars are also called dwarfs (never "dwarves"). A main sequence star is a dwarf even if it has ten times more mass than the Sun.

TECHNICAL STUFF

When astronomers and science writers refer to "normal stars," they often mean main sequence stars. When they write about "Sun-like stars," they may mean main sequence stars with roughly the same mass as the Sun, give or take a factor of no more than two. The writers may also be distinguishing stars on the main sequence, no matter how massive, from stars like white dwarfs and neutron stars.

The smallest main sequence stars — much less massive than the Sun — are *red dwarfs,* which shine with a dull red glow. Red dwarfs have little mass, but they exist in enormous quantities. The vast majority of main sequence stars are red dwarfs. Like tiny gnats at the seashore, they float all around you, but you can hardly see them. Red dwarfs are so dim that you can't see even the nearest one, Proxima Centauri — which is, in fact, the nearest-known star of any kind beyond the Sun — without telescopic aid. For more on Proxima, see Chapter 14.

Red dwarfs are so much smaller, less massive, and fainter than stars like the Sun that you may be tempted to ignore them. But as mentioned earlier in this chapter, a red dwarf lasts forever, while more massive stars, like the Sun, eventually die. We may be proud of our Sun, but those puny red dwarfs will have the last laugh.

Red giants: Burning out the golden years

Red giant stars represent another kettle of starfish entirely. Red giants are much larger than the Sun, but someday the Sun will become a red giant itself (see Chapter 10). Often they measure as big around at the equator as the orbit of Venus or even the orbit of Earth.

The giants represent a stage in the life of a star after the main sequence, at least for stars with between a few times less and a few times more than the mass of the Sun. Aldebaran in the constellation Taurus and Arcturus in Bootes are red giants that are easily seen with the naked eye. I list both of them in Chapter 1.

A typical red giant doesn't burn hydrogen in its core; in fact, it burns hydrogen in a spherical region just outside the core, called a *hydrogen-burning shell*. A red giant can't burn hydrogen in its core because it has already turned all of its core hydrogen into helium through nuclear fusion. (Some red giants generate their energy in other ways, but they're less common.)

Stars much more massive than the Sun don't become red giants; they swell up so much that astronomers call them *red supergiants*. A typical red supergiant can be 1,000 or 2,000 times larger than the Sun and big enough to extend past the orbit of Jupiter, or even Saturn, if put in the Sun's place. Take a look at Betelgeuse in Orion or Antares in Scorpius; they're two of the brightest stars in the sky and they're both red supergiants.

Closing time: Coming up on the tail end of stellar evolution

The *end states of stellar evolution* is a catchall term for stars whose best years are far behind them. The category encompasses

>> Central stars of planetary nebulae

>> White dwarfs

>> Supernovas

>> Neutron stars

>> Black holes

These objects are all dying stars on their final glide paths, doomed to oblivion, or the corpses of stars that used to be.

THE BIGGEST STARS ARE THE LONELIEST

SETI observers (the Search for Extraterrestrial Intelligence — see Chapter 14) don't point their radio telescopes at massive stars to search for radio signals from advanced civilizations. Why not? Because massive stars explode and die after lifetimes so short that scientists can't imagine intelligent life (or even primitive life) evolving on any surrounding planets before the end comes.

Massive stars are much rarer than low-mass stars. The more massive the stars, the fewer they are. So eventually, as existing stars age and the birth clouds for new stars are used up, the Milky Way will consist overwhelmingly of just two types of stars: the red dwarfs that go on more or less forever and the white dwarfs that fade as they go. Yes, neutron stars and stellar mass black holes will dot the Milky Way, but because they represent the remains of the much rarer, more massive stars, they'll be numerically insignificant compared to the red and white dwarfs, which come from the most abundant types of main sequence stars.

Stars are like people in that the biggest ones are rare, just as 7-foot, 5-inch basketball players are few and far between.

Central stars of planetary nebulae

Central stars of planetary nebulae are little stars at the centers (duh!) of a certain type of small, beautiful nebulae. (You can see a photo in the color section of this book.) These nebulae have nothing to do with planets, but in early telescopes, their images resembled greenish planets like Uranus — hence the name.

Central stars of planetary nebulae are like white dwarfs — in fact, they turn into white dwarfs. So the central stars, too, are the remains of Sun-like stars. The nebulae, each composed of gas that a star expelled over tens of thousands of years, expand, fade, and blow away. Eventually, they leave behind stars that no longer serve as the centers of anything — they become white dwarfs. These fading stars would be prime candidates for the reality show circuit.

White dwarfs

White dwarfs can actually be white, yellow, or even red, depending on how hot they are. White dwarfs are the remains of Sun-like stars that take after the old generals who, according to Douglas MacArthur, never die — they just fade away.

A white dwarf is like a glowing coal from a freshly extinguished fire. It doesn't burn anymore, but it still gives off heat. White dwarfs are the most common stars after red dwarfs, but even the closest white dwarf to Earth is too dim to see without a telescope.

White dwarfs are compact stars — small and very dense. A typical white dwarf may have as much mass as the Sun, yet it takes up as much space as Earth, or a little bit more. So much matter is packed into such a small space that a teaspoon of a white dwarf would weigh about a ton on Earth. Don't try measuring it with your good silver; the spoon will get all bent out of shape.

Supernovas

Supernovas (which experts call *supernovae,* as though they all studied Latin like old-time scientists) are enormous explosions, some of which destroy entire stars (see Figure 11-2). Various kinds of supernovas exist, but I introduce you to just the two principal varieties.

FIGURE 11-2:
A supernova (the bright star in the bottom left corner) in the spiral galaxy M74.

Courtesy of ESO/PESSTO/S. Smartt

The first type you need to know about is Type II (hey, I didn't invent the numbering system). A *Type II supernova* is the brilliant, catastrophic explosion of a star much larger, brighter, and more massive than the Sun. Before the star exploded, it was a red supergiant or maybe was even hot enough to be a blue supergiant. Regardless of color, when a supergiant explodes, it may leave behind a little souvenir, which is a neutron star. Or much of the star may implode (fall in on its own center) so effectively that it leaves behind an even weirder object, a black hole.

The second important type of supernova is called a Type Ia. *Type Ia supernovas* are even brighter than Type IIs, and they explode in a reliable manner. The actual brightness, or luminosity, of a Type Ia is always about the same; therefore, when astronomers observe a Type Ia supernova, we can figure out how far away it is by how bright it appears to us on Earth. The farther away it is, the dimmer the supernova looks. Astronomers use Type Ia supernovas to measure the universe and its expansion. In 1998, two groups of astronomers studying Type Ia supernovas discovered that the expansion of the universe isn't slowing down — it's expanding at an ever-faster rate. This discovery was contrary to previous belief, so it led experts to revise their theories of cosmology and the Big Bang and to recognize the existence of the mysterious *dark energy*. (I describe dark energy and the Big Bang in Chapter 16.)

Type Ia supernovas all produce similar explosions because they erupt in binary systems (covered later in this chapter) in which gas from one star flows down onto the other (a white dwarf), building up an outer hot layer that reaches a *critical mass* and then explodes, shattering the white dwarf. Unlike a Type II supernova, which may leave a neutron star or a black hole in place of the exploded star, a Type Ia leaves nothing but an expanding gas cloud. With less than critical mass, no explosion occurs. With critical mass, a standard explosion results. With more than the critical mass . . . wait — you can't have more than critical mass because the star already exploded! Astrophysics isn't so hard, is it?

Experts have argued for years over the kind of binary system that produces a Type Ia supernova. According to one theory, the two stars in the binary are a white dwarf and a companion, possibly like the Sun. The white dwarf sucks gas off its bigger partner. Another promising theory asserts that both stars in the binary system are white dwarfs. Both theories may even be correct: Some Type Ia supernovae may come from a big star/little star kind of system, and others may come from a pair of equals.

Neutron stars

Neutron stars are so small that they look up to white dwarfs, but they also outweigh them. (More accurately, they "outmass" them. Weight is just the force that a planet or other body exerts on an object of a given mass. You'd weigh different amounts on the Moon, Mars, and Jupiter than you do on Earth, even though your mass stays the same.)

Neutron stars are like Napoleon: small in stature, but not to be underestimated. (Figure 11-3 features a neutron star.) A typical neutron star spans only one or two dozen miles across, but it has half again or even twice the mass of the Sun. A teaspoon of neutron star material would weigh about a billion tons on Earth.

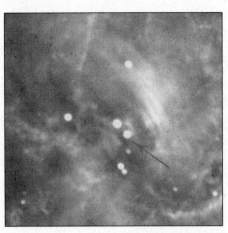

FIGURE 11-3:
A pulsar
(at the arrow) in
the center of the
Crab Nebula.

Some neutron stars are better known as *pulsars*. A pulsar is a highly magnetized, rapidly rotating neutron star that produces one or more beams of radiation (which may consist of radio waves, X-rays, gamma rays, and/or visible light). As a beam sweeps past Earth like a searchlight from a galactic supermarket opening, our telescopes receive brief spurts of radiation, which we call "pulses." So guess how pulsars get their name. Your pulse rate tells you how rapidly your heart is beating. A pulsar's rate tells how fast it turns. The rate can be as much as 700 hundred times per second or just once every few seconds. That makes my head spin just thinking about it.

Black holes

Black holes are objects so dense and compact that they make neutron stars and white dwarfs seem like cotton candy. So much matter is packed into such a small space in a black hole that its gravity is strong enough to prevent anything, even a ray of light, from escaping. Physicists theorize that the contents of a black hole in effect have left our universe. If you fall into a black hole, you can kiss your universe good-bye.

You can't see the light from a black hole because the light can't get out, but scientists can detect black holes by their effects on surrounding objects. Matter in the vicinity of a black hole gets hot and rushes madly around, but it never gets organized; instead, the powerful gravity of the black hole pulls most of it into the hole and "that's all folks."

Actually, I oversimplified; some of the matter swirling around the black hole does escape — just in time sometimes. It shoots it out in powerful jets at a significant fraction of the speed of light (which is 186,000 miles per second in a vacuum such as outer space).

This is how scientists detect black holes:

>> Gas swirls around the black hole, spiraling inward, in a flattened cloud called an *accretion disk;* as it gets closer to the black hole, it gets very hot and emits visible light, X-rays, or other radiation, which scientists observe.

>> Radio telescopes or other instruments pick up those jets of high-energy particles I mention earlier. They come from the accretion disk or another region very close to but outside the black hole.

>> Stars race around orbits at fantastic speeds, driven by the gravitational pull of an enormous unseen mass.

>> Gravitational waves from the collision of two black holes sweep out through the universe, and the Laser Interferometer Gravitational-Wave Observatory, with its detectors in Louisiana and Washington State, detects them.

Astronomers have found lots of evidence for two basic kinds of black holes and limited but increasing information on a third kind. Here's a rundown of the three types:

>> **Stellar mass black holes:** These black holes have — you guessed it — the mass of a star. More accurately, these black holes range from about three times the mass of the Sun up to perhaps a hundred times the solar mass, although astronomers haven't found any as heavy as that. Stellar mass black holes are about the size of neutron stars. A black hole with ten times the mass of the Sun has a diameter of about 37 miles (60 kilometers). If you could squeeze the Sun down to a size compact enough to make it a black hole (fortunately, this is probably impossible), its diameter would be 3.7 miles (6 kilometers). Stellar mass black holes form in supernova explosions and possibly by other means.

>> **Supermassive black holes:** These monsters have masses of hundreds of thousands to more than 20 billion times the mass of the Sun (for some examples, see Chapter 13). Generally, supermassive black holes are located at the centers of galaxies. I want to say that they "gravitate" there, but probably most of them form right there, or the galaxy forms around them. The Milky Way Galaxy has a central black hole, known as Sagittarius A*. (Nope, the asterisk doesn't refer you to a footnote. When saying the name out loud, you say *Sagittarius A star.*) It weighs in at about 4 million solar masses, and we in the solar system orbit around that black hole once every 226 million years — a period of time astronomers call the galactic year. Astronomers think that a supermassive black hole exists at the center of all or most galaxies. But we're not so sure about the smallest ones, called dwarf galaxies. (Get the scoop on galaxies in Chapter 12.)

TECHNICAL STUFF

When I talk about the size of a black hole, I mean the diameter of its *event horizon*. The event horizon is the spherical surface around the black hole where the velocity needed for something to escape the black hole equals the velocity of light. Outside the event horizon, the escape velocity is smaller, so light or even high-speed matter can escape. All the material inside the event horizon is crushed and compacted into a tiny, dense region in the center.

>> **Intermediate mass black holes (IMBHs):** IMBHs are a poorly understood class of black holes, meaning that astronomers don't know what they are or whether they really exist. They have estimated masses that range from about one hundred to about ten thousand times the mass of the Sun. An IMBH is more massive than any known star, so it probably wasn't formed by the collapse of a single star (which is how stellar mass black holes are formed). On the other hand, IMBHs are found outside the central regions of galaxies, while supermassive black holes are found in galactic centers. So IMBHs aren't made where supermassive black holes form, and they aren't made in the collapse of a star, as is a stellar mass black hole. What creates an IMBH? Inquiring minds want to know, but I wouldn't look for the answer in the *National Enquirer.*

To tell the truth, supermassive black holes aren't stars. Most likely, neither are intermediate mass black holes. But I have to mention them someplace! You can't call yourself an astronomer if you don't know about black holes. (Check out Chapter 13 for even more about them.) If you pass yourself off as an astronomer, folks will ask you all kinds of questions about black holes. But how many questions do you think you'll get about main sequence stars and young stellar objects?

Star Color, Brightness, and Mass

The significance of the different types of stars (see the section "Life Cycles of the Hot and Massive") becomes clearer when you see basic observational data plotted on an astrophysicist's graph. The data are the magnitudes (or brightnesses) of the stars, which are plotted on the vertical axis, and the colors (or temperature), which are marked on the horizontal axis. This graph is called a *color-magnitude diagram,* or the Hertzsprung-Russell or H-R diagram after the two astronomers who first made such diagrams (see Figure 11-4).

As a college astronomy teacher, I could always tell who studied and who didn't. When I asked what data are plotted on the H-R diagram, the students who answered "H and R" revealed that they were just guessing.

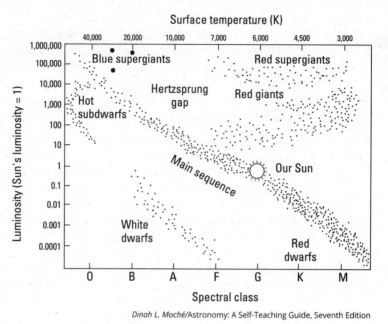

Surface temperature (K)

FIGURE 11-4:
The Hertzsprung-
Russell diagram
shows star
brightness and
temperature.

Dinah L. Moché/Astronomy: A Self-Teaching Guide, Seventh Edition

Spectral types: What color is my star?

Hertzsprung and Russell didn't have good information on the colors or temperatures of stars, so they plotted the spectral type on the horizontal axis of their original diagrams. The *spectral type* is a parameter assigned to a star on the basis of its spectrum. The *spectrum* is the way the light of a star appears when a prism or other optical device in an instrument called a spectrograph spreads it out.

**TECHNICAL
STUFF**

At first, astronomers didn't have a clue about what spectral types meant, so one of them, Williamina Fleming, just grouped stars together (under the headings of Type A, Type B, and so on) based on similarities in their spectra. Later, one of Fleming's colleagues, Annie Jump Cannon, simplified the system and dropped some of the original types while also rearranging them. The spectral types she retained ended up reflecting the temperatures and other physical conditions in the atmospheres of the stars, where their light emerges into space. It was only later that scientists understood what the colors signified and that the spectral types were organized in temperature order.

The main spectral types on the H-R diagram are O, B, A, F, G, K, and M, going from the hottest stars to the coolest. College students memorize this sequence with the help of mnemonic devices such as "Oh, be a fine girl (guy), kiss me." Table 11-1 describes the general properties of stars of each spectral type.

TABLE 11-1 **Spectral Types of Stars**

Type	Color	Surface Temperature	Example
O	Violet-white	32,000°K or more	Lambda Orionis
B	Blue-white	10,000°K to 32,000°K	Rigel
A	White	7,300°K to 10,000°K	Sirius
F	Yellow-white	6,000°K to 7,300°K	Procyon
G	Yellower-white	5,300°K to 6,000°K	Sun
K	Orange	3,900°K to 5,300°K	Arcturus
M	Red	Less than 3,900°K	Antares

Star light, star bright: Luminosity classifications

The spectral types O, B, A, F, G, K, and M have subdivisions denoted by Arabic numerals, which allow astronomers to classify stars in finer detail. Each spectral letter type has up to ten subdivisions. For example G stars encompass ten types, from G0 to G9. The lower the subdivision number, the hotter the star. The Sun's spectral type is G2, while Beta Aquilae (listed in Table 11-2) has spectral type G8. In other words, the Sun is hotter than Beta Aquilae, and the latter is almost cool enough to fall in spectral type K.

I tell you all this so that when you look up a star like the Sun or Beta Aquilae in an astronomy book (or other resource, such as the Internet), you know what it means when it lists the Sun as G2 or Beta Aquilae as G8.

But wait, there's more: Some books just use G2 or G8 for these two stars, but others describe the Sun and other stars with an additional tag, usually a Roman numeral. So you may see the Sun called a G2 V star and Beta Aquilae referred to as G8 IV. Astronomers call the Roman numeral the star's *luminosity class*.

Spectral type, like G2, refers to the temperature of a star, while luminosity class, like IV or V, describes its size and also its average density (because larger stars usually have lower density than small stars). I summarize stars by luminosity class and size in Table 11-2.

TABLE 11-2 **Luminosity Classes for Stars**

Class	Description	Example
I	Supergiant	Rigel
II	Bright giant	Gamma Aquilae
III	Giant	Aldebaran
IV	Subgiant	Beta Aquilae
V	Main sequence (dwarf)	Rigil Kentaurus
D	White dwarf	Sirius B

You'll occasionally find a star listed as luminosity class Ia or Ib. These designations give astronomers more detail: *Ia* refers to a brighter supergiant, and *Ib* indicates a fainter supergiant. But any supergiant is much brighter than the stars of the other luminosity classes.

TECHNICAL STUFF

D, the luminosity class for white dwarfs, could have been defined as the Roman numeral *D*, meaning 500. But it's actually just the English letter *D*, short for "dwarf." After you've mastered the subject of spectral types and luminosity classes, you can flash a *V* for victory, but to an astrophysicist, it's also a Roman numeral *V* for main sequence dwarf.

The brighter they burn, the bigger they swell: Mass determines class

A star with greater mass concentrates a fiercer nuclear fire in its core and produces more energy than a star of lower mass, so a more massive main sequence star is brighter and hotter than a less massive main sequence star. The more massive stars usually are bigger, too. With this information, you can follow the fundamental point of stellar astrophysics reflected in the H–R diagram: Mass determines class.

On the H–R diagram (see Figure 11-4), magnitude (or luminosity) is plotted with greater luminosities or brighter magnitudes higher on the graph, and spectral type is plotted with the hotter stars to the left and the cooler stars to the right. Temperature runs from right to left, and magnitude runs from top to bottom.

On any H-R diagram, the plot of real observational data, where each point represents a single star, reveals a great deal to the careful reader:

>> Most stars are on a band that runs diagonally from upper left to lower right. The diagonal band represents the main sequence, and all the stars on the band are normal stars like the Sun, burning hydrogen in their cores.

>> Some stars are on a wider, sparser, and roughly vertical band that stretches up and a little to the right from the diagonal band toward brighter magnitudes (higher luminosities) and cooler temperatures. This band is the *giant sequence* and consists of red giant stars.

>> A few stars are located all across the top of the diagram, from left to right. These stars are supergiants; blue supergiants are on the left side of the diagram, more or less, and red supergiants (which outnumber the blue) are on the right side.

>> A few stars are located far below the diagonal band, down at the bottom left and bottom center of the diagram. These stars are white dwarfs.

Astronomers plot a main sequence star on the H-R diagram according to its brightness and temperature, but its brightness and temperature depend on only its mass. The diagonal shape of the main sequence represents a trend from high-mass to low-mass stars. The stars on the upper left of the main sequence have higher masses than the Sun, and the stars on the far right have lower masses.

Astronomers usually don't plot young stellar objects on the same H-R diagram with all the other stars, but if they did, they'd plot the YSOs on the right side of the diagram, above the main sequence — but not nearly as high up as the supergiant stars. Neutron stars and black holes (which are invisible!) are too dim to plot on the same H-R diagrams with normal stars.

The H-R diagram

With just a little more explanation, you, too, can be a stellar astrophysicist and understand in one fell swoop why stars fall into different parts of the H-R diagram. Researchers spent decades figuring this out, but I give it to you on a plate. To keep it simple, I discuss a calibrated H-R diagram, where all the stars are plotted according to true brightness.

Consider this: Why is one star brighter or dimmer than another? Two simple factors determine the brightness of a star: temperature and surface area. The bigger the star, the more surface area it has, and every square inch (or square centimeter) of the surface produces light. The more square inches, the more light. But what about the amount of light a given square inch produces? Hot objects burn brighter than cool objects, so the hotter the star, the more light it generates per square inch of surface area.

Got all that? Here's how it all goes together:

» **White dwarfs** are near the bottom of the diagram because of their small size. With very few square inches of surface area (compared to stars like the Sun), white dwarfs just don't shine as brightly. As they fade away like old generals, they move down the diagram (because they get dimmer) and farther to the right (because they get cooler). You don't see many white dwarfs on the right side of the H-R diagram because the cool white dwarfs are so faint that they usually fall below the bottom of the diagram as printed in books.

» **Supergiants** are near the top of the H-R diagram because they're super big. A red supergiant can be more than 1,000 times larger than the Sun (if you put a supergiant in the Sun's place, it would extend beyond Jupiter's orbit). With all that surface area, supergiants are naturally very bright.

The fact that the supergiants are roughly at the same height on the diagram from left to right indicates that the blue supergiants (the ones on the left) are smaller than the red supergiants (the ones on the right). How do you know that? Supergiants are blue because they burn hotter, and if they burn hotter, they produce more light per square inch. Because their magnitudes are roughly the same (all the supergiants are near the top of the diagram), the red supergiants must have larger surface areas to produce equal total light (because they produce less light from each square inch).

» **Main sequence** stars are on the diagonal band that runs from the upper left to the lower right on the diagram, because the main sequence consists of all stars that burn hydrogen in their cores, regardless of their size. But a difference in size affects where the main sequence stars appear on the H-R diagram. The hotter main sequence stars (the ones on the left) are also bigger than the cool main sequence stars, so hot main sequence stars have two things going for them: They have larger surface areas, and they produce more light per square inch than the cool stars. The main sequence stars at the far right are the dim and cool red dwarfs.

THE BROWN DWARFS DON'T TOP THE CHARTS

Brown dwarfs — discovered in the mid-1990s — are recent additions to the celestial inventory. They're smaller and less massive than stars and about as big as a gas giant planet like Jupiter, but they're much more massive than Jupiter. (Brown dwarf masses range from about 13 to about 70 times the mass of Jupiter.) They shine by their own light, like a star, not by reflected light, like Jupiter. But brown dwarfs aren't true stars because nuclear fusion operates only briefly in their cores. After fusion stops, they generate no more energy and just cool and fade. Their spectral types range from the cool end of type M to the successively cooler types, L and T. (Astronomers suspect that there may be even cooler brown dwarfs; if so, they'd belong to spectral class Y.) On the H-R diagram in Figure 11-4, brown dwarfs fall at the extreme bottom right or even just off the chart to the right of the bottom-right corner. NASA is looking for citizen scientists to help it find nearby brown dwarfs. If you want more info on joining that project, check out Chapter 9.

Eternal Partners: Binary and Multiple Stars

Two stars or three or more stars that orbit around a common center of mass are called binary stars or multiple stars, respectively. Studies of binary and multiple stars help scientists understand how stars evolve. These small stellar systems are also fun to observe with backyard telescopes.

Binary stars and the Doppler effect

About half of all stars come in pairs. These *binary stars* almost always are *coeval*, a fancy term for "born together." Stars that form together, united by their mutual gravity as they condense from their birth clouds, usually stay together. What gravity unites, other celestial forces rarely can break apart. A grown star in a binary system has never had another partner (well, hardly ever — some strange cases occur in dense star clusters, where stars can come so close that they may actually lose or gain a partner).

A binary system consists of two stars that each orbit a common *center of mass.* The center of mass of two stars that have exactly equal masses falls exactly halfway between them. But if one star has twice the mass of the other star, the center of mass is closer to the heavier star; in fact, the center is twice as far from the lighter star as from the heavier star. If one star has one-third the mass of its heavy companion, it orbits three times farther from the center of mass, and so on. The two stars are like kids on a seesaw: The heavier kid has to sit closer to the pivot so the two are in balance.

The two stars in a binary system follow orbits equal in size if the stars are equal in mass. Stars with different masses follow orbits of different sizes. The general rule: The big guys follow smaller orbits. You may think that binary systems are like our solar system, where the closer a planet orbits to the Sun, the faster it goes and the less time it takes to make one complete orbit. That idea may be a reasonable one, but it's wrong nonetheless.

In binary systems, the big star that follows the smaller orbit travels slower than the little star in a big orbit. In fact, their respective speeds depend on their respective masses. The star that carries one-third the mass of its companion moves three times as fast. By measuring orbital velocity, astronomers can determine the relative masses of a binary system's members.

The fact that the orbital speeds of the member stars of a binary system depend on mass is what makes binary stars attract the high interest of astronomers. If one star is three times more massive than the other, it moves around its orbit in the binary star system at one-third the orbital speed of its companion star. All astronomers have to do to figure out the stars' relative masses (meaning how much more massive one star is than the other) is measure their velocities. However, only rarely can astronomers track the stars as they move because most binary stars are so far away that we can't see them moving around their orbits. Fortunately, instead of throwing up their hands in defeat, astronomers have been able to measure star masses by studying the light from a binary star and analyzing its spectrum — a spectrum that may be the combined light of both stars in the binary.

A phenomenon called the *Doppler effect* helps astronomers figure out the masses of binary stars by studying their stellar spectra. Here's all you need to know about the Doppler effect, named for Christian Doppler, a 19th-century Austrian physicist:

The frequency or wavelength of sound or light, as detected by the observer, changes, depending on the speed of the emitting source with respect to the observer. For sound, the emitting source may be a train whistle. For light, the source may be a star. (Higher-frequency sounds have a higher pitch; a soprano has a higher pitch than a tenor.) Higher-frequency light waves have shorter wavelengths, and lower-frequency light waves have longer wavelengths. In the simple case of visible light, the shorter wavelengths are blue and the longest wavelengths are red.

According to the Doppler effect,

>> If the source is moving toward you, the frequency that you detect or measure gets higher, so

  • The pitch of the train whistle seems to be higher.

  • The light from the star seems to be bluer.

>> If the source is moving away, the frequency gets lower, so

  • The whistle you hear has a lower pitch.

  • The star looks redder.

The train whistle is the official example of the Doppler effect that instructors have taught to generations of sometimes unwilling high school and college students. But who listens to train whistles anymore? The same effect happens when an emergency responder speeds past you with its siren blaring. The next time you see flashing red and blue lights, listen for the siren to drop in pitch as it passes you by. Another analogy is the way you feel the waves on the water as you zip around in a motorboat. As you ride toward the direction from which the waves are coming, you feel the boat rocked rapidly by choppy waves. But when you head back toward the beach, the rocking becomes much gentler and the same waves are less choppy. In the first case, you moved toward the waves, meeting them before they would've met you if you stood (or floated) still. So the frequency at which the waves struck the boat was greater than if the boat had remained at rest. The frequency of the waves doesn't change, but the frequency of the waves that you sense changes.

The spectrum of a star contains some dark lines — places (wavelengths or colors of light) where the star doesn't emit as much light as at adjacent wavelengths. The decreased emission at those wavelengths is caused by the absorption of light by particular kinds of atoms in the atmosphere of the star. The dark lines form recognizable patterns, and when the star is moving back and forth in its orbit, the Doppler effect makes the patterns of lines move back and forth in the spectrum detected on Earth. When the spectral lines shift toward longer wavelengths, the phenomenon is a *redshift.* When they shift toward shorter wavelengths, it's a *blueshift.* Other ways of producing redshifts and blueshifts exist, but the Doppler effect is the most familiar cause.

So by observing the spectra of binary stars and seeing how their spectral lines shift from red to blue to red again as the stars orbit, astronomers can determine their velocities and, therefore, their relative mass. And by seeing how long it takes for a spectral line to go as far to the red as it goes and then how long it takes to go as far to the blue as it goes and back to the red again, astronomers can tell the duration or period of the binary star orbit.

If you know that the period of one complete orbit is 60 days, for example, and you know how fast the star is moving, you can figure out the circumference of the orbit and, thus, the radius of its orbit. After all, if you drive nonstop from New York City to a town in upstate New York in three hours (good luck with the traffic!) at 60 miles per hour, you know that the distance you travel is 3×60, or 180 miles.

Two stars are binary, but three's a crowd: Multiple stars

Double stars are two stars that appear close to each other as seen from Earth. Some double stars are true binaries, orbiting their common centers of mass. But others are just *optical doubles,* or two stars that happen to be in nearly the same direction from Earth but at very different distances. They have no relation to each other; they haven't even been introduced.

Triple stars are three stars that appear to be close but, like the members of a double star, may or may not be all that close. But a *triple-star system,* like a binary system, consists of three stars held together by their mutual gravitation that all orbit a common center of gravity.

A comparison to wedded (or unwedded) bliss may be in order. "Three's a crowd" is an expression of the instability in most romantic arrangements when a third person becomes involved. The same is true of triple-star systems: They consist of a close pair or binary system and a third star in a much bigger orbit. If all three stars moved on close orbits, they'd interact gravitationally in chaotic ways, and the group would break up, with at least one star flying away, never to return. So a triple system is effectively a "binary star" in which one member is actually a very close star-pair.

Quadruple stars are often "double-doubles," consisting of two close binary star systems that each revolve around the common center of mass of the four stars.

Multiple stars is the collective term for star systems larger than binaries: triples, quadruples, and more. At some point, the distinction between a large multiple-star system and a small star cluster becomes blurred. One is essentially the same as the other. (I describe star clusters in Chapter 12.)

STELLAR SPECTROSCOPY IN A NUTSHELL

Stellar spectroscopy is the analysis of the lines in the spectra of stars and is by far an astronomer's most important tool for investigating the physical nature of stars. Spectroscopy reveals

- Radial velocities (motions toward or away from Earth) of stars

- Relative masses, orbital periods, and orbit sizes of stars in binary systems

- Temperatures, atmospheric densities, and surface gravities of stars

- Magnetic fields and their strengths on stars

- Chemical composition of stars (what atoms are present and in what states they exist)

- Sunspot cycles of stars (well, starspot cycles)

All this information comes from measuring the positions, widths, and strengths (how dark or how bright they are) of the little dark (or sometimes bright) lines in the spectra of stars. Scientists analyze them with the help of the Doppler effect to find out how fast stars move, the size of their orbits, and their relative masses. Other phenomena, such as the *Zeeman effect* and the *Stark effect*, affect the appearance of the spectral lines. Applying their knowledge of these effects, astronomers can figure the strength of the magnetic field of the star from the Zeeman effect and determine the density and surface gravity in the star's atmosphere from the Stark effect. The very presence of particular spectral lines, each of which comes from a specific kind of atom that's absorbing (dark lines) or emitting (bright lines) light in the atmosphere of a star, tells astronomers about some of the chemical elements present in the atmosphere of the star and the temperatures in the star where those atoms are emitting or absorbing the light.

The spectral lines even tell astronomers what condition or *ionization state* the atoms are in. Stars are so hot that the heat may strip atoms of iron, for example, of one or more of their electrons, making them iron ions. Each type of iron ion, depending on how many electrons it has lost, produces spectral lines with different characteristic patterns and positions in the spectrum. By comparing the spectra of stars recorded with telescopes with the spectra of chemical elements and ions as measured in laboratory experiments or calculated with computers, astronomers can analyze a star without ever coming within light-years of it.

In cool stellar gases, much of the iron loses only one electron per atom, so it produces the spectrum of singly ionized iron. But in the very hot parts of stars, such as the million-degree corona of the Sun, iron may lose ten electrons; the element is in a

(continued)

(continued)

high-ionization state, and it produces the corresponding pattern of spectral lines. That pattern clearly points to the existence of a very high temperature region on the star.

Certain parts of the Sun's spectrum change along with the coming and going of disturbed regions on the Sun, which peak about every 11 years (as I explain in Chapter 10). Similar changes occur in the spectra of other Sun-like stars. So astronomers can tell the length of the sunspot cycle of a distant star by using spectroscopy, even though the star is too far away to catch a glimpse of its sunspots.

Change Is Good: Variable Stars

Not every star is, as Shakespeare wrote, as "constant as the Northern Star." In fact, the North Star isn't constant, either. The famous star, also called Polaris, is a *variable star,* meaning one whose brightness changes from time to time. For many years, astronomers thought that they had the North Star's brightness changes down pat. It seemed to brighten a little and fade a little over and over, reproducibly. But suddenly its expected changes, well, changed. This difference in the pattern may signify a physical change over time, and scientists are studying what it means. Recently, astronomers at Villanova University concluded that the North Star has brightened by about 1 magnitude (about 2.5 times) since antiquity.

Variable stars come in two basic types:

>> **Intrinsic variable stars:** These stars change in brightness because of physical changes in the stars themselves. These stars divide into three main categories:

- Pulsating stars
- Flare stars
- Exploding stars

>> **Extrinsic variable stars:** These stars seem to change in brightness because something outside the star alters its light, as visible from Earth. The two main types of extrinsic variable stars are

- Eclipsing binaries
- Microlensing event stars

I describe each of these basic types of variable stars in the following section.

Go the distance: Pulsating stars

Pulsating stars bulge in and out, getting bigger and smaller, hotter and cooler, brighter and dimmer. These stars simply oscillate like throbbing hearts in the sky.

Cepheid variable stars

The most important pulsating stars, from a scientific standpoint, are the Cepheid variable stars, named for the first studied star of their type, Delta in the constellation Cepheus (Delta Cephei).

American astronomer Henrietta Leavitt discovered that Cepheids have a *period-luminosity relation.* This term means that the longer the period of variation (the interval between successive peaks in brightness), the greater the true average brightness of the star. So an astronomer who measures the apparent magnitude of a Cepheid variable star as it changes over days and weeks, and who thereby determines the period of variability, can readily deduce the true brightness of the star.

Why do astronomers care? Well, knowing the true brightness enables us to determine the distance of the star. After all, the farther the star, the dimmer it looks, but it still has the same true brightness.

TECHNICAL STUFF

Distance dims stars according to the *inverse square law:* When a star is twice as far away, it looks 4 times as faint; when the distance is tripled, the star looks 9 times as faint; and when a star is 10 times farther away, it looks 100 times as faint.

The headlines about the Hubble Space Telescope determining the distance scale and age of the universe came from a Hubble study of Cepheid variable stars. Those Cepheids are in faraway galaxies. By tracking their brightness changes and by using the period-luminosity relation, the Hubble observers figured out just how far away the galaxies are.

Henrietta Leavitt's period–luminosity relation made that Hubble Space Telescope work possible. Earlier, Edwin Hubble used her discovery to learn that the universe is expanding (see Chapter 16). For her work at Harvard College Observatory, Leavitt earned 30 cents an hour when not volunteering, about half the pay a male colleague doing the same work would've earned at the time.

RR Lyrae stars

RR Lyrae stars are similar to Cepheids, but not as big and bright. Some RR Lyrae stars are located in globular star clusters in our Milky Way, and they have a period-luminosity relation, too.

Globular clusters are huge balls of old stars that were born while the Milky Way was still forming. A few hundred thousand to a million or so stars are all packed in a region of space only 60 to 100 light-years across. Observing the changes in brightness of RR Lyrae stars enables astronomers to estimate their distances, and when the stars are in globular clusters, it tells us how far away the clusters are. (For more on globular and other star clusters, turn to Chapter 12.)

Why is it so important to know the distance of a star cluster? Here goes: All the stars in a single cluster were born from a common cloud at the same time, and they're all at nearly the same distance from Earth because they exist in the same cluster. So when scientists plot the H–R diagram of stars in a cluster, the diagram is free of errors that may be caused by differences in the distances of the stars. And if scientists know the distance of the cluster, they can convert all the plotted magnitudes to actual luminosities, or the rates at which the stars produce energy per second. Such quantities can be directly compared with astrophysical theories of the stars and how they generate their energy. This stuff keeps astrophysicists busy.

Long period variable stars

Astrophysicists celebrate the information gleaned from Cepheid and RR Lyrae variable stars. Amateur astronomers, on the other hand, delight in observing long period variables, also called Mira stars, Miras, or Mira variables. Mira is the popular name for the star Omicron Ceti, in the constellation Cetus (the Whale), the first-known long period variable star.

Mira variables are huge red stars that pulsate like Cepheids, but they have much longer periods, averaging ten months or more, and the amount by which their visible light changes is even greater. At its brightest, Mira itself is visible with the naked eye, and at its faintest, you need a telescope to spot it. The changes of a long period variable star are also much more variable than the changes of a Cepheid. The brightest magnitude that a particular long period variable star reaches can be quite different from one period to the next. Such changes are easily observed, and they constitute basic scientific information. You can participate in these and other variable star studies, as I explain in the later section "How to Help Scientists by Observing the Stars."

Explosive neighbors: Flare stars

Flare stars are little red dwarfs that suffer big explosions, like ultrapowerful solar flares. You can't see most solar flares without the aid of special colored filters because the light of the flare is just a tiny fraction of the total light of the Sun. Only the rare, very large "white light" flares are visible on the Sun without a special filter. (But you still need to use one of the techniques for safe solar viewing that I describe in Chapter 10.) However, the explosions on flare stars are so bright that the magnitude of the star changes detectably. You're looking at the star through a telescope, and suddenly it shines brighter. Not all red dwarfs have these frequent explosions, but Proxima Centauri, a red dwarf and the nearest star beyond our Sun, is a flare star.

Nice to nova: Exploding stars

The explosions of novas and supernovas are so large that I can't lump them in with the flare stars; they're enormously more powerful and have much greater effects.

Novas

A nova explodes through a build-up process on a white dwarf in a binary system, much like the Type Ia supernova explosions that I describe earlier in this chapter. But in a Type Ia supernova, the white dwarf is shattered; in a nova, the white dwarf isn't destroyed. The star just blows its top, and then it settles down, sucking more gas off its companion and onto its surface layer. The powerful gravity of the white dwarf compresses and heats this layer, and after centuries or millennia, off it goes again.

At least, that's the theory. No scientist has been around long enough to see an ordinary or *classical nova* explode twice. But similar binary systems exist in which the explosions aren't quite as powerful as in a classical nova but recur frequently enough that amateur astronomers are always monitoring them, ready to announce the discovery of a new explosion and alert professionals to study it. These objects have various names, including *dwarf nova* and *AM Herculis systems*. Classical novas, dwarf novas, and similar objects are known collectively as *cataclysmic variables.*

A nova bright enough to see with the naked eye occurs about once a decade, give or take. I studied one in Hercules for my doctoral thesis in 1963. If it hadn't exploded at just the right time, I may have waited ten years for a thesis topic. Recently, bright novas in Scorpius, Delphinus, Centaurus, and Sagittarius delighted astronomers in 2007, 2013, 2013, and 2015, respectively.

Red novas

Recently astronomers observed several examples of a new kind of stellar explosion, *red novas.* The observations showed that they are more luminous than novas and other cataclysmic variables that I mention in the preceding section but less powerful than supernovas, which I cover in the next section. They are caused by a different process than those other explosions. Experts suspect that red novas occur when two main sequence stars in a closely spaced binary star spiral in toward each other, merge, and explode. As you can guess from the name, they look red.

You may get to see a red nova with the naked eye. A team of astronomers who study a binary star called KIC 9832227 predict that it will explode as a red nova in 2022. They think it will be visible without a telescope, in the Northern Cross asterism in constellation Cygnus (I define asterisms in Chapter 1). To find the Northern Cross for the first time and to see a suitable sky map of it, head to www.constellation-guide.com/northern-cross/. If the prediction comes true, a red star will pop up in the Northern Cross in about 2022, but it may come a year or so earlier or later. If the prediction doesn't pan out, some astronomers' faces will be red.

Supernovas

Supernovas throw off nebulae, called *supernova remnants,* at high speeds (see Figure 11-5). The nebula at first consists of the material that made up the shattered star, minus any remaining central object, be it neutron star or black hole (see the section "Closing time: Coming up on the tail end of stellar evolution" earlier in this chapter). But as it expands into space, the nebula sweeps up interstellar gas like a snowplow that accumulates snow. After a few thousand years, the supernova remnant is mostly swept-up gas rather than supernova debris.

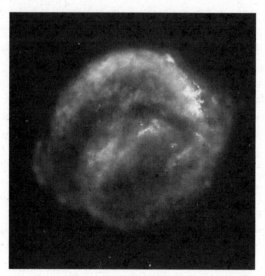

FIGURE 11-5:
The Kepler Supernova Remnant is 14 light-years wide.

Courtesy of NASA/ESA/Johns Hopkins University

Supernovas are incredibly bright and rather rare. Astronomers estimate that, in a galaxy like the Milky Way, a supernova occurs every 25 to 100 years, but we haven't witnessed a supernova in our home galaxy since Kepler's Star in 1604, before the invention of the telescope. Others may have occurred, obscured from view by dust clouds in the Galaxy. A huge southern star known as Eta Carinae looks like it may be on the verge of going supernova in the Milky Way, but in astronomers' parlance, that means it may explode at any time within the next million years.

Only one supernova has been visible with the naked eye since 1604. It was Supernova 1987A, located in our neighbor galaxy, the Large Magellanic Cloud, or LMC (which I describe in Chapter 12). The supernova was too far south to be seen from the continental United States, but I wasn't going to miss a celestial event of such great rarity, so I flew to Chile to see it. The Chilean astronomers gave me a warm reception.

Hypernovas

Hypernovas are especially bright supernovas that seem to produce at least some of the gamma ray bursts that flash in the sky from time to time. The bursts are extremely powerful blasts of high-energy radiation emitted in beams like the beams from searchlights. NASA launched the Swift satellite in November 2004 to discover more about them. When Swift detects a burst coming from a certain direction, it, well, swiftly notifies observatories on the ground to focus on that part of the sky. Hypernovas are much rarer than other supernovas, and no hypernova has been seen in our own Galaxy.

TIP

If you want to know more about Swift and its discoveries, go to NASA's Swift website at swift.gsfc.nasa.gov/ and also check out the Swift Education and Public Outreach site at swift.sonoma.edu. If you have an iPad or an iPhone, download the free Swift Explorer App — it has some cool capabilities.

Stellar hide-and-seek: Eclipsing binary stars

Eclipsing binary stars are binary systems whose true brightness doesn't change (unless one of the two stars happens to be a pulsating star, flare star, or other intrinsic variable), even though they look like variable stars to us. The *orbital plane* of the system — the plane that contains the orbits of the two stars — is oriented so that it contains our line of sight to the binary system. So on every orbit, one star eclipses the other, as seen from Earth, and the star's brightness dips during the eclipse. (Of course, the situation reverses halfway through the orbital period, when the eclipsed star now does the eclipsing.)

If the two stars in a binary system have orbital periods of four days, then every four days, the more massive star in the system, usually called *A*, passes exactly in front of the other star in the view from Earth. This pass blocks all or most of the light from star *B* from reaching Earth (depending on its size compared to *A* — sometimes the less massive star is larger than its heavy companion), so the binary looks fainter. Astronomers call this event a *stellar eclipse.* Two days after the eclipse, star *B* passes in front of star *A,* creating another eclipse.

In the earlier section "Binary stars and the Doppler effect," I mention how astronomers use the orbital velocities to figure out the relative masses of the stars. Well, they can also use the velocities to find the diameters of stars. Scientists take spectra and discover how fast the stars orbit by using the Doppler effect, and they measure the durations of the eclipses in eclipsing binaries. A stellar eclipse of star *B* begins when the leading edge of *A* starts to pass in front of it. The eclipse ends when the following edge of *A* finishes passing in front of *B*. So the orbital speed multiplied by the duration of the eclipse tells scientists how big star *A* is.

In all these methods, the fine details are a little more complicated, but you can easily understand the principles of stellar investigation.

The most famous eclipsing binary is Beta Persei, also known as Algol, the Demon Star. You won't have a devil of a time observing Algol's eclipses in the Northern Hemisphere — Algol is a bright star well placed for observation in the northern sky in the fall. You can watch its eclipses without a telescope or even binoculars. Every 2 days and 21 hours, Algol's brightness dims by a little more than one magnitude — more than a factor of 2.5 — for about two hours. But you need to know *when* to look for an eclipse. You don't want to stand around in the backyard for almost three days. The neighbors will talk. Check out the pages of *Sky & Telescope* that list information for observers. Usually, you find a paragraph called "Minima of Algol," which lists the dates and times when eclipses will occur during a period of a month or two. (If you don't see a list in the current issue, it means that Algol is too close to the Sun in the sky for observation that month.) The eclipse times are stated in Universal Time (UT). You can easily convert them to your local standard time or daylight savings time, as I explain in Chapter 5.

Minima are the times when variable stars, extrinsic or intrinsic, reach the lowest brightness levels of their ongoing cycles. *Maxima* are the times when the stars shine brightest.

Hog the starlight: Microlensing events

Sometimes a faraway star passes precisely in front of an even farther star. The two stars are unrelated and may be thousands of light-years from each other, but the gravity of the nearer star bends the paths taken by light rays from the farther star in such a way that the distant star appears much brighter from Earth for a few days or weeks. This effect is predicted from Einstein's Theory of General Relativity, and astronomers regularly detect it. When the gravity of a huge object like a galaxy bends light, astronomers call the process *gravitational lensing.* When the gravity of a body as small as a star bends light, it's called *microlensing.*

You may be thinking how unlikely it is that two unrelated stars would line up perfectly with Earth, and you'd be right! Congratulations on that thought. To detect such a rare event on a regular basis, astronomers use telescopic electronic cameras that can record hundreds of thousands to millions of stars at a time. With that many stars under observation, a foreground star passes in front of a distant star every so often, even though astronomers don't know beforehand which two stars are involved.

The trick is to point the electronic camera at a region where you can see vast numbers of stars simultaneously in the field of view. Such regions include the Large Magellanic Cloud, a nearby satellite galaxy of the Milky Way (see Chapter 12), and the central bulge of the Milky Way itself, where a whole mess of stars hang out.

Your Stellar Neighbors

When you look at Alpha Centauri with the naked eye, you see one bright star. When you check it out with a telescope, you see two bright stars close together in the field of view. The two stars form a binary system. But a third star, Proxima Centauri, makes it a triple system. You don't see Proxima in the telescope's field of view with the two bright stars because it's in a huge orbit around them and, as seen from Earth, is more than 2° away, or more than four times the apparent diameter of the full Moon. (Note that Proxima isn't visible to the naked eye. I discuss it in the earlier section "Main sequence stars: Enjoying a long adulthood;" I also describe Proxima's planet and a possible interstellar mission to fly past it in Chapter 14.)

For a complete look at the triple system, check out the following list:

>> **Alpha Centauri (also called Rigil Kentaurus):** This star is a bright, G-type star in the southern constellation Centaurus (see Figure 11-6). It's a main sequence dwarf with about the same color as the Sun but is somewhat brighter.

>> **Alpha Centauri B:** Alpha Centauri's orange companion is a slightly smaller and cooler main sequence dwarf.

>> **Alpha Centauri C:** Our nearest stellar neighbor beyond the Sun, the little red dwarf and flare star, is also called Proxima Centauri.

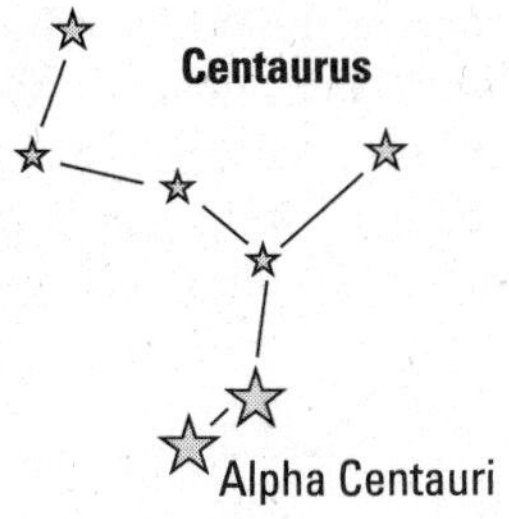

FIGURE 11-6:
Alpha Centauri is a triple-star system in the far southern sky.

© John Wiley & Sons, Inc.

The Alpha Centauri system is about 4.4 light-years from Earth, with Proxima on the near side, at 4.2 light-years. The system is in the far southern sky, so you need to be in the Southern Hemisphere, or at least very far south in the Northern Hemisphere, to view it.

Sirius, at a distance of 8.6 light-years, is the brightest star in the night sky. It's also known as Alpha Canis Majoris, in Canis Major (the Great Dog; see Figure 11-7). Slightly south of the Celestial Equator, Sirius is easily visible from most inhabited places on Earth. It's a white, A-type, main sequence star that shines bright enough to make folks ask, "What's that big star?"

Like most stars other than the Sun, Sirius has a companion star: Sirius B, a white dwarf. Sirius is known as the Dog Star, and when the American telescope maker Alvan Clark discovered its tiny companion, Sirius B, in 1862, naturally someone nicknamed it "the Pup." Sirius and Sirius B orbit their center of mass once every 50 years. (I define center of mass in the earlier section "Binary Stars and the Doppler effect.")

Vega is Alpha Lyrae, the brightest star in the constellation Lyra (the Lyre). It appears high in the sky at temperate northern latitudes (such as the U.S. mainland) on summer evenings and is an object that most amateur astronomers know like the back of their hands. Located about 25 light-years from Earth, Vega is an A-type main sequence star like Sirius. Vega is a brilliant white sparkler and one of the most conspicuous stars in the sky.

Betelgeuse isn't really in the solar neighborhood; it's a red supergiant star of spectral type M, about 640 light-years from Earth. But everybody likes its name, which many pronounce "Beetle Juice" (which is as good a way to say it as any), and observers enjoy its deep red color. It is, after all, a red supergiant, more than 20,000 times brighter than the Sun. Although Betelgeuse is Alpha Orionis, the brightest star in Orion is actually Rigel (Beta Orionis), a blue-white supergiant about 860 light-years from us.

How to Help Scientists by Observing the Stars

Thousands of stars are under watch because they vary in brightness or exhibit some other special characteristic. Professional astronomers can't keep up with them all, and that's where you come in. You can monitor some stars with your own eyes, with binoculars, or with a telescope.

You need to be able to recognize the stars and judge their magnitudes. The brightness of many stars changes so significantly — by a factor of two, ten, or even hundreds — that eye estimates are sufficiently accurate to keep track of them. The trick is to use a *comparison chart,* a map of the sky that shows the position of the variable star and the positions and magnitudes of *comparison stars.* A comparison star has a known brightness that doesn't vary (or so we hope).

TIP

The American Association of Variable Star Observers (AAVSO) offers a wealth of information explaining how to observe variable stars. Check out its website at www.aavso.org. It provides what you need to start watching variable stars. You can observe some bright variable stars with the naked eye and more of them with binoculars. When you use a good small telescope, the sky's the limit!

AAVSO encourages beginners and expert amateur observers alike. You can download its *Manual for Visual Observing of Variable Stars* in your choice of English or various other European, Asian, and Middle Eastern languages. (Amateur astronomers around the world contribute their observations of variable stars to AAVSO's research.)

Check out AAVSO's Variable Star Plotter (VSP) on its website. You can enter the name or number of a variable star, and VSP will create a sky chart for the star that you can download for use with your telescope. After you've read the manual and practiced judging star magnitudes, you'll be all set to observe the variable star and send your findings to AAVSO.

Many amateurs enjoy observing Mira, the famous star I describe in the earlier section "Long period variable stars." Mira and similar stars have such great brightness changes that a star you see near its maximum brightness may get too faint to find with your telescope as it approaches minimum brightness. Fortunately, you can see a yearly table of predicted dates for the maximum phases of a selection of these stars at a web page of the British Astronomical Society. Look it up at www.britastro.org/vss/mira_predictions.htm.

Star Studies to Aid with Your Brain and Computer

If you live in a place where weather conditions and/or urban lighting aren't conducive to astronomical observations on a frequent basis, you can still help astronomers by practicing *citizen science:* You use your home computer to examine telescopic data from orbiting spacecraft or from ground-based professional observatories, in accordance with online instructions. All you need to do is sign up on a project website, study the project tutorial, and begin examining the data.

Thousands of interested people enroll in these projects, so although any one individual may not have the best scientific judgment, the reports from many participants average out. The citizen scientist findings generally alert professional astronomers to interesting and possibly scientifically important objects or phenomena in the (literally) astronomical databases that no one expert can scan by himself.

Here are two good citizen science projects to consider:

» **The Milky Way Project (`www.zooniverse.org/projects/povich/milky-way-project`):** This program, led by California State Polytechnic University, Pomona, seeks new information on how stars form.

  You examine images from two NASA spacecraft, the Spitzer Space Telescope and the Wide-field Infrared Survey Explorer (WISE). When you participate, you use the computer tools provided on the project site to draw circles around objects that the project staff call green bubbles and red or orange arcs in the Milky Way (examples are given so you know what to look for). Professional astronomers use the information you submit to benefit ongoing research on star formation in the Milky Way. (I describe the Milky Way in Chapter 12.)

» **Disk Detective (`www.diskdetective.org/`):** This project also uses images from the WISE spacecraft. You scan them to help the scientists find circumstellar disks associated with young stellar objects, which I describe in the earlier section "Young stellar objects: Taking baby steps." I hope you step to up to the task.

Chapter **12**

Galaxies: The Milky Way and Beyond

Our solar system is a tiny part of the Milky Way galaxy, a great system of a few hundred billion stars, thousands of nebulae, and hundreds of star clusters. The Milky Way, in turn, is one of the largest components of the Local Group of Galaxies. Beyond the Local Group is the Virgo Cluster, the nearest large cluster of galaxies — about 54 million light-years from Earth. As scientists peer out into the universe over much greater distances, they see *superclusters*, immense systems that contain many individual clusters of galaxies. So far, we haven't found superclusters of superclusters, but *Great Walls*, which are immensely long superclusters, do exist. And much of the universe seems to consist of gigantic cosmic voids, which contain relatively few detectable galaxies.

This chapter introduces you to the Milky Way and its most important parts, and takes you farther into the universe to meet other types of galaxies.

Unwrapping the Milky Way

The Milky Way, also called "the Galaxy," is a lot bigger than the candy bar, if not as sweet. But it does have a creamy-looking center. The Milky Way is seen from

Earth as the wide band of diffuse light that you can view best on clear summer and winter nights from a dark location.

A stream of milk through the universe was as good as any explanation for the Milky Way until 1610, when Galileo first observed it with a telescope. He found that the Milky Way is nothing to lap about: It consists of a huge number of stars, most of them so faint or so far away that they blend into one large fuzzy region on the sky. Most of the stars in the Milky Way are invisible to the eye, but as a group, they shine. Obviously, the telescope was a definite improvement for studying the Milky Way (and almost everything else in astronomy)!

Galaxies are the basic building blocks of the universe, and the Milky Way is a good-size block. It contains almost everything in the sky that you can see with the naked eye — from Earth and our solar system to the stars of the solar neighborhood, the visible stars in the constellations, and all the stars that blend together to make that milky stream in the night sky — and plenty of objects and matter that you can't see. It also contains nearly every nebula you can see without telescopic aid — and plenty more, to boot.

Now that's a tall glass of milk! Besides loose stars, the Milky Way holds hundreds of star clusters, such as the Pleiades and Hyades in the constellation Taurus and, for lucky viewers in Australia, South America, and other points far south, the Jewel Box in Crux, the (Southern) Cross, and the magnificent star cluster Omega Centauri.

UNCLOAKING THE MURKY MILKY WAY

In the past, stargazers viewed the Milky Way with ease, but now many people can't see it or don't know that it exists because they live in or near cities with so many lights that the sky is bright with light pollution rather than dark as nature intended.

The solution? Flee the light pollution, at least once in awhile. Go out to the mountains or the shore on your vacation or on a weekend and check out a darker sky than you have at home. The light of the full Moon interferes with seeing the Milky Way, too, so plan your jaunt around the time of the new Moon, when there is little or no moonlight in the sky. The Milky Way is most prominent in the sky during summer and winter, and least visible in spring and fall. (For tips on dark-sky sites for stargazing, turn to Chapter 2; for more on this subject and a lot of information about light pollution and its harmful effects, visit the International Dark-Sky Association website at www.darksky.org.)

How and when did the Milky Way form?

The age of the universe is about 13.8 billion years; the estimated ages of the oldest known stars in the Milky Way are over 13 billion years too. Thus the Milky Way is nearly as old as the universe. (I describe the origin of the universe according to current theory in Chapter 16.)

Here's the short story of the Milky Way: Long ago, gravity caused a gigantic cloud of primordial gas to fall together and condense. As little clumps inside the cloud collapsed even faster than the cloud as a whole, stars formed. Although the big cloud must have turned very slowly at first, it would've rotated faster and faster as it became smaller and flattened into the present-day spiral disk structure. And before you knew it, *voilà, la voie lactee* (French for "there it is, the Milky Way"). Actually, its formation isn't quite that simple because the Milky Way is a glutton — it has swallowed smaller neighbor galaxies for eons, adding their stars to its own collection. It continues its feast today. What a menace!

You just read my favorite Milky Way theory, bar none. If you have a better one, become an astronomer yourself and write your own book someday — in science, theories make the world go 'round, and maybe even the Galaxy.

What shape is the Milky Way?

The Milky Way is the shape it is and the size it is because, in the universe, gravity rules. The Milky Way is a spiral galaxy, consisting of a pizza-shape formation of billions of stars (the *galactic disk,* about 100,000 light-years in diameter) that contains the spiral arms (see Figure 12-1). The arms are shaped roughly like the streams of water from a rotating lawn sprinkler and contain many bright, young blue and white stars and gas clouds. Groups of young, hot stars (called *associations*) dot the spiral arms in the galactic disk like pepperoni slices on a pizza. Bright and dark nebulae seem to mushroom all around the arms, along with huge molecular clouds, such as Monoceros R2 (its location is marked in Figure 12-1), where most of the gas is cool and dim. Between the arms are the *interarm regions* (not all astronomical terms are as catchy as Barnacle Bill, the name of a rock on Mars, or the Red Rectangle, a nebula shaped like an hourglass — go figure).

At the center of our galaxy is a place called (you guessed it!) the *galactic center.* Surrounding the center is the galactic bulge, which puts a sumo wrestler to shame. The *galactic bulge* is a roughly spherical formation of millions of mostly orange and red stars, sitting like a great meatball at the center of the galactic disk and extending far above and below it. But at least some of the bulge stars are arranged in an elongated formation, more sausage-shaped than meatball-shaped. Astronomers call the sausage a bar. When a spiral galaxy has a prominent bar, it's called a *barred spiral* (I discuss barred spirals later in this chapter), but the bar in the Milky Way isn't very conspicuous.

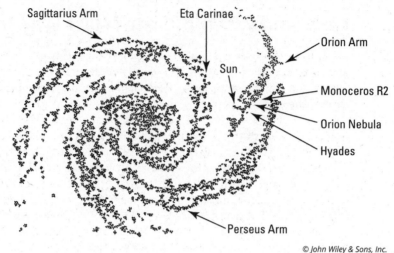

Sagittarius Arm

Eta Carinae

Orion Arm

Sun

Monoceros R2

Orion Nebula

Hyades

Perseus Arm

© John Wiley & Sons, Inc.

FIGURE 12-1:
The Milky Way is
a spiral galaxy
with arms
encircling the
galactic center.

At the center of the galactic bulge is Sagittarius A*, a supermassive black hole. Figure 12-1 presents a model of the Milky Way with its toppings and ingredients. (It's a close-up of the galactic disk, with the galactic bulge omitted for clarity.) Any way you slice it, the Milky Way is a terrific galaxy.

The flat imaginary surface or midplane of the galactic disk is called the *galactic plane,* and the circle that represents its intersection with the sky, as visible from Earth, is called the *galactic equator.*

Sometimes astronomers list the position of an object in galactic coordinates rather than in right ascension and declination (coordinates that I define in Chapter 1). The galactic coordinates are *Galactic Latitude,* which is measured in degrees north or south of the galactic equator, and *Galactic Longitude,* which is measured in degrees along the galactic equator.

Galactic Longitude starts at the direction to the galactic center, which is 0° longitude. (Actually, the zero point of Galactic Longitude is slightly off the galactic center because scientists placed it where the galactic center was thought to be in 1959; we know better now.) Galactic Longitude proceeds along the galactic equator from the constellation Sagittarius into Aquila, Cygnus, and Cassiopeia. It goes on through Auriga, Canis Major, Carina, and Centaurus, and all the way to 360° longitude, back at the galactic center. When you look with binoculars at the constellations I just named, you see more stars, star clusters, and nebulae than elsewhere in the sky. The "plane truth" is that the constellations that the galactic plane intersects are among the finest sights in the sky.

TIP

Check out the map of the Milky Way based on measurements of more than one billion stars by the Gaia satellite. The European Space Agency publicly released it in 2016. Just surf to `www.esa.int/Our_Activities/Space_Science/Gaia/Gaia_s_billion-star_map_hints_at_treasures_to_come`. That may be an awfully long web address, but think how long the list of the stars is.

TIP

You can find panoramic maps of the Milky Way's galactic plane — as recorded with radio telescopes, X-ray and gamma-ray observatory satellites, and visible light (or "optical") telescopes on the ground — at NASA's MultiWavelength Milky Way website at `mwmw.gsfc.nasa.gov`.

Where can you find the Milky Way?

The Milky Way isn't a certain distance from Earth; it contains our planet. The galactic center is about 27,000 light-years from Earth. Our solar system takes about 226 million years to make one orbit around the center. This period is the galactic year. Be glad you don't have to wait that long for your next birthday to come around.

Stars closer to the galactic center than the Sun take less time to go around the center, and stars farther from the center take more than our 226 million years.

The outskirts of the galactic disk — or *galactic rim*, as science fiction fans know it — is, at its closest part to Earth, about an equal distance in the opposite direction from the galactic center. The disk of the Milky Way is pretty much identical to the milky band of light in the sky.

SEEING BEYOND THE MILKY WAY

The three objects that are beyond the Milky Way but readily visible with the naked eye are the Large and Small Magellanic Clouds (two nearby galaxies visible from the Southern Hemisphere) and the Andromeda Galaxy. Some people blessed with excellent vision (and many others who want to impress their friends) say that they can see the Triangulum Galaxy, too. Both the Andromeda and Triangulum galaxies are between 2 and 3 million light-years from Earth, but Andromeda is bigger and brighter.

I count the Large Magellanic Cloud as one object, but it actually contains a huge bright nebula, the Tarantula, which you can make out with the naked eye. (Don't worry, it won't bite.) For a few months in 1987, you could see a bright supernova in the Large Magellanic Cloud near the Tarantula, Supernova 1987A. It was the first supernova visible to the naked eye since Kepler's Star in 1604, although it wasn't in our own galaxy like Kepler's Star. Unlike Kepler's Star, Supernova 1987A wasn't visible from Europe or the continental United States, but was readily seen from places like Australia, Chile, and South Africa.

The Milky Way is about 163,000 light-years from a galaxy called the Large Magellanic Cloud, about 2.5 million light-years from the Andromeda Galaxy, and about 54 million light-years from the nearest big cluster of galaxies, the Virgo Cluster. It also falls smack dab in the middle of a little cluster of galaxies (sizes are relative here), the Local Group.

Star Clusters: Meeting Galactic Associates

Star clusters are bunches of stars located in and around a galaxy. They haven't associated by chance (even though one type of star cluster is called an *association*); they're groups of stars that formed together from a common cloud and, in many cases, are held together by gravity. The three main types of star clusters are *open clusters, globular clusters,* and *OB associations.*

TIP

For superb images of star clusters, head to the European Southern Observatory website (www.eso.org); click on "Images" and then on "Categories" and "Star Clusters." You see beautiful color photos that seem unlabeled, but move your cursor over an image, and the caption appears. Then click on the picture, and a new page appears with a larger version of the photo, more details about the pictured star cluster, and links that you click for a free download of the picture in your choice of format.

You can also check out a photo of a star cluster in the color section of this book.

A loose fit: Open clusters

Open clusters contain dozens to thousands of stars, have no particular shape, and are located in the disk of the Milky Way. Typical open clusters span about 30 light-years. They aren't highly concentrated (if at all) toward their centers, unlike globular clusters (see the next section), and open clusters typically are much younger. Open clusters are great for viewing with small telescopes and binoculars, and you can see some with the naked eye. You can find them marked on most good star atlases, such as *Sky & Telescope's Pocket Sky Atlas* by Roger W. Sinnott (Sky Publishing). In this atlas, open clusters are marked by yellow disks with dotted edges, and the disk sizes are proportional to the apparent sizes of the clusters as seen from Earth. (The atlas also shows globular clusters, which I describe in the next section.)

The most famous and easily seen open clusters in the northern sky are as follows:

>> **The Pleiades (also known as the Seven Sisters):** Located in the northwest corner of Taurus, the Bull, the Pleiades looks to the naked eye like a tiny dipper. You can compare your eyesight with a friend's by seeing how many stars you can count in the Pleiades, which is M45, the 45th object in the *Messier Catalog* (see Chapter 1). Gaze through binoculars and see how many more you can find. The brightest star in the Pleiades is Eta Tauri (3rd magnitude), also called Alcyone. (See Chapter 1 for an explanation of magnitudes.) The Japanese name for the Pleiades is *Subaru;* next time you see a Subaru in the parking lot, take a close look at its emblem.

>> **The Hyades:** The Hyades is also located in Taurus and makes another great sight for the naked eye. It includes most of the stars that make up the V shape in the head of Taurus. You can't miss this cluster because the V features the bright red giant Aldebaran (1st magnitude), which is Alpha Tauri (see Figure 12-2). Aldebaran is known as the Bull's Eye and perhaps explains why the bull's-eye of a dartboard is often colored red. Aldebaran actually is far beyond the Hyades cluster, but it appears in the same direction from Earth.

The Hyades looks much bigger than the Pleiades because it's about 150 light-years from Earth versus about 450 light-years for the Pleiades.

>> **The Double Cluster:** Located in Perseus, the Hero, the Double Cluster is a beautiful sight through binoculars and especially through a small telescope. Its two clusters are NGC 869 and 884, each roughly 7,600 light-years from Earth. NGC stands for *New General Catalogue,* which was new when it first appeared in 1888; it doesn't list any generals (or captains or colonels, for that matter).

>> **The Beehive (also called Praesepe):** The Beehive (Messier 44) is the main attraction of Cancer, the Crab, a constellation composed of dim stars. It looks like a nice fuzzy patch with the naked eye and like a swarm of many stars when seen through binoculars. This cluster is about 600 light-years from Earth.

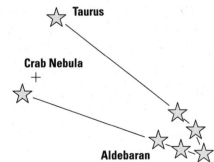

FIGURE 12-2: Taurus contains the red giant Aldebaran (Alpha Tauri).

© John Wiley & Sons, Inc.

Viewers in the Southern Hemisphere can enjoy these open clusters:

>> **NGC 6231:** In Scorpius, the Scorpion, NGC 6231 is a southern-sky object, but you can readily see it from much of the United States in the evening during the summer. You need to be in a dark place with an unobstructed southern horizon. The observer Robert Burnham Jr. (who also wrote the classic three-volume *Celestial Handbook*) described it as resembling "a handful of diamonds on black velvet."

>> **NGC 4755 (also called the Jewel Box):** Located in Crux, the Cross, the Jewel Box includes the bright star Kappa Crucis. Crux, popularly known as the Southern Cross, is a perennial favorite of viewers in the Southern Hemisphere. If you take a cruise through the South Seas, insist that the ship have an astronomy lecturer on board. (I will probably be available.) He or she can point out the Southern Cross; with binoculars, you can enjoy the fine sight of the Jewel Box. Your view won't match the photo of the Jewel Box from the Very Large Telescope at Paranal Observatory in Chile (www.eso.org/public/images/eso0940a/), but there's a thrill in seeing the cluster in person.

A tight squeeze: Globular clusters

Globular clusters are the retirement homes of the Milky Way Galaxy. They appear to be about as ancient as the galaxy itself (some experts think that the globular clusters were the first objects to form in the Milky Way), so they contain ancient stars, including many red giants and white dwarfs (see Chapter 11). The stars you can see in a globular cluster with your telescope are mostly red giants. With bigger telescopes, you can see orange and red main sequence dwarf stars. Only the Hubble Space Telescope and other very powerful instruments can pick out many of the much fainter white dwarfs in a globular cluster.

A typical globular star cluster contains a hundred thousand to a million or more stars, all packed in a ball (hence the term *globular*) that measures just 60 to 100 light-years in diameter. The closer the stars are to the center, the more tightly they're packed in (see Figure 12-3). Its high degree of concentration and its great number of stars distinguish a globular cluster from an open cluster.

Another key difference is that open clusters are distributed across the galactic disk in a great flat pattern, but globular clusters are arranged spherically around the center of the Milky Way. Most globular clusters are concentrated toward the galactic center, but many of the globulars you can easily see are well above or below the galactic plane.

FIGURE 12-3:
Globular cluster
Messier 4 in
Scorpius.

Here are the best globular clusters for viewing in the northern sky:

>> **Messier 13:** This globular cluster makes its home in Hercules, representing the mythical character of the same name.

>> **Messier 15:** You can find this globular cluster in Pegasus, the Winged Horse.

You can spot both M13 and M15 with the naked eye under suitably dark sky conditions, but you may need to reassure yourself with binoculars or a small telescope, which show the clusters as fuzzy spots larger than stars. Use a star chart (such as *Sky & Telescope's Pocket Sky Atlas*, which I mention in the preceding section) to locate the clusters.

Observers in the Northern Hemisphere are cheated out of the best globular star clusters because by far the two biggest and brightest shine in the deep southern sky:

>> **Omega Centauri:** Located in Centaurus, the Centaur

>> **47 Tucanae:** In Tucana, the Toucan

These clusters are spectacular sights through small binoculars and just about worth the trip to South America, South Africa, Australia, or other places where you can readily see them. Bring a friend because two can share a view of the Toucan.

Be sure to check out the photo of a globular cluster in the color section.

Fun while it lasted: OB associations

OB associations are loose stellar groupings with dozens of stars of spectral types O and B (the hottest types of main sequence stars) and sometimes fainter, cooler stars (see Chapter 11 for more about spectral types). Unlike with globular clusters and large open clusters, gravity doesn't hold together OB associations; over time, the stars move away from each other, dissolving the association like a limited partnership that is breaking up. OB associations are located close to the galactic plane.

Many of the bright young stars in the constellation Orion are members of the Orion OB association. (See Chapter 3 for more about Orion.)

Taking a Shine to Nebulae

A nebula is a cloud of gas and dust in space ("dust" meaning microscopic solid particles, which may be made of silicate rock, carbon, ice, or various combinations of those substances; and "gas" meaning hydrogen, helium, oxygen, nitrogen, and more, but mostly hydrogen). As I note in Chapter 11, some nebulae play an important role in star formation; others form from stars gasping on their deathbeds. Between the cradle and the grave, nebulae come in a number of varieties. (Check out a photo of a nebula in the color section of this book.)

Here are a few of the most familiar nebulae:

>> **H II regions** are nebulae in which hydrogen is ionized, meaning that the hydrogen loses its electron. (A hydrogen atom has one proton and one electron.) The gas in an H II region is hot, ionized, and glowing, due to the effects of ultraviolet radiation from nearby O or B stars. All the large, bright nebulae that you can see through binoculars are H II regions. (*H II* refers to the ionized state of the hydrogen in the nebula.) These nebulae often appear red or pink in colored images. The Lagoon Nebula in Sagittarius is a good example.

>> **Dark nebulae,** also known as H I regions, are the dust bunnies of the Milky Way, consisting of clouds of gas and dust that don't shine. Their hydrogen is neutral, meaning that the hydrogen atoms haven't lost their electrons. The term *H I region* refers to the neutral (non-ionized) state of the hydrogen. Dark nebulae are most easily visible against a bright background, where they have plenty of light to block out and make their presence known. The Horsehead Nebula in Orion is one of the most famous dark nebulae.

>> **Reflection nebulae** consist of dust and cool, neutral hydrogen. They shine by the reflected light of nearby stars. Without the nearby stars, they'd be dark nebulae.

Sometimes a new reflection nebula appears suddenly, and you may discover it, as amateur astronomer Jay McNeil did. In January 2004, he found a new reflection nebula in the constellation Orion with a 3-inch refractor in his backyard, and professionals now call it McNeil's Nebula. But don't hold your breath; this type of discovery is rare. Reflection nebulae often appear blue in colored images, as you can see when you examine images of the Pleiades star cluster in Taurus and its famous reflection nebula.

>> **Giant molecular clouds** are the largest objects in the Milky Way, but they're cold and dark, and scientists would've gazed right by them if not for the data gathered by radio telescopes, which can detect emissions of faint radio waves from molecules in these clouds, such as carbon monoxide (CO). Like all other nebulae, giant molecular clouds are mostly made of hydrogen, but scientists often study them by means of their trace gases, such as CO. The hydrogen in giant clouds is molecular, with the designation H_2, which means that each molecule consists of two neutral hydrogen atoms.

One of the most exciting nebular discoveries of the 20th century was that bright H II regions, such as the Orion Nebula, are hot spots on the peripheries of unseen giant molecular clouds. For centuries, people could see the Orion Nebula but didn't know that it's no more than a bright pimple on a huge invisible object, the Orion Molecular Cloud. According to current ideas, new stars are born in molecular clouds, and when they get hot enough, they ionize their immediate surroundings, turning them into H II regions. The part of a molecular cloud where the dust is thick enough to cut off the light of many or most of the stars behind the cloud, as visible from Earth, is called a dark nebula.

H II regions, dark nebulae, giant molecular clouds, and many of the reflection nebulae are located in or near the Milky Way's galactic disk. Two other interesting types of nebulae are planetary nebulae and supernova remnants, which I cover briefly in the following sections (and in Chapter 11).

Picking out planetary nebulae

Planetary nebulae are produced by old stars that started out resembling the Sun but then expelled their outer atmospheric layers. (The Sun will do the same thing in the far future; see Chapter 10). The expelled gases make up the nebulae, which are ionized and made to glow by ultraviolet light from the hot little stars at their centers. The tiny stars are all that remain of the former suns. Planetary nebulae expand into space and fade as they grow larger. They can be well off the galactic plane, unlike H II regions. (See a planetary nebula in the color section of this book.)

For decades, astronomers believed that many or most planetary nebulae were roughly spherical. But now we know that most of these nebulae are *bipolar,* meaning that they consist of two round lobes projecting from opposite sides of the central star. The planetary nebulae that look spherical, such as the Ring Nebula in the constellation Lyra (see Figure 12-4), are bipolar, too, but the axis down the center of the lobes happens to point toward Earth (and so, like a dumbbell viewed end on, they look circular). Astronomers took many years to figure this out, so maybe we're dumbbells, too.

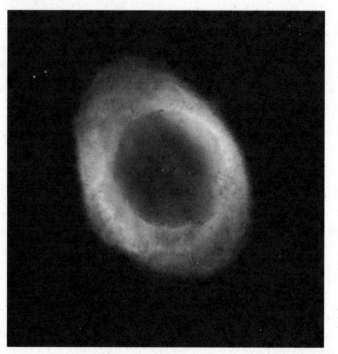

FIGURE 12-4:
The Ring Nebula in Lyra is bipolar but appears spherical from Earth.

Courtesy of NASA

CORRECTING A GALACTIC GOOF-UP

Until the 1950s, astronomers used the term *nebula* to refer to a galaxy because, until the 1920s, they thought galaxies other than the Milky Way were nebulae in the Milky Way. In other words, astronomers believed that there was only one galaxy: the Milky Way.

It took a few dozen years for the change in understanding to prevail in the language of astronomy. Therefore, the authors of astronomy books have only recently stopped referring to the Andromeda Galaxy as the Andromeda Nebula.

Edwin P. Hubble, for whom the Hubble Space Telescope is named, wrote the famous book *The Realm of the Nebulae*. The book was all about galaxies, not nebulae as astronomers use the term now. Among his achievements, Hubble proved that the Andromeda Nebula is a galaxy full of stars, not a big cloud of gas. A former boxer, he fought in World War I, smoked a pipe, and supposedly bullied other astronomers at Mount Wilson Observatory, but his discoveries are no bull.

TECHNICAL STUFF

Curious point: *Protoplanetary nebulae* are much studied by astrophysicists, but there are actually two kinds, and they have nothing to do with each other. One type of protoplanetary nebula is the early stage of a planetary nebula — a phase in the death of a star (not to be confused with the *Star Wars* Death Star). The other type of protoplanetary nebula is the birth cloud of a solar system's star and its planets. Yes, astronomers use the same term to refer to two completely different kinds of objects, but nobody's perfect.

Breezing through supernova remnants

Supernova remnants begin as material ejected from massive stellar explosions. A young supernova remnant consists almost exclusively of the shattered remains of the exploded star that expelled it. But as the gas moves outward through interstellar space, it resembles a rolling stone that *does* gather moss. The expanding remnant creates a snowplow effect as it pushes along (see Chapter 11) and accumulates the thin gas of interstellar space. After some thousands of years, the remnant is overwhelmingly composed of this "plowed up" interstellar gas, and the remains of the exploded star are mere traces. Supernova remnants are located along or near the galactic plane of the Milky Way.

Enjoying Earth's best nebular views

TIP

Nebulae are among the most beautiful sights through small telescopes. You need a good star chart, like those found in *Sky & Telescope's Pocket Sky Atlas,* and you want to start with an easy target, like the Orion Nebula, which you glimpse by eye and binoculars before homing in on it with your telescope. For H II regions like the Orion Nebula, a telescope with a low f/number, such as the Orion ShortTube 80-A Equatorial Refractor, may work the best (see Chapter 4, where I discuss using the scope for hunting comets, for more info on this particular tool). For smaller nebulae like the Ring Nebula, which I describe in the following list, the Meade ETX90 Observer telescope (see Chapter 3) is a good choice for a beginner. The Meade has a computerized control system that points the telescope accurately at objects that you can't see with the naked eye.

The following are some of the best, brightest (or, for dark nebulae, darkest), and most beautiful nebulae you can see from northern latitudes, including some southern-sky objects that aren't located very far south of the celestial equator:

>> **The Orion Nebula,** Messier 42 (see Chapter 1), in Orion, the Hunter

   You can easily see the Orion Nebula, an H II region, with the naked eye as a fuzzy spot in the sword of Orion. It looks fine through binoculars and spectacular through a small telescope. The telescope also shows the Trapezium, a bright quadruple star (see Chapter 11) in the nebula.

>> **The Ring Nebula,** Messier 57, in Lyra

   The Ring Nebula is a planetary nebula high in the sky at northern temperate latitudes on a summer evening. As with all planetary nebulae, you need to use a star chart to find it with your telescope, unless you have a computer-assisted telescope such as the Meade ETX-90 (see Chapter 3), which points right to the nebula at your command.

>> **The Dumbbell Nebula,** Messier 27, in Vulpecula, the Fox

   The Dumbbell Nebula, along with the Ring Nebula, is among the easiest planetary nebulae to spot with a small telescope. The best time for observation is the summer and fall.

>> **The Crab Nebula,** Messier 1, in Taurus

   The Crab Nebula is the remains of a supernova that exploded in the year 1054, as seen from Earth and recorded by Chinese astronomers. It appears as a fuzzy spot through a small telescope, but a big telescope shows two stars near its center. One star isn't associated with the Crab; it just appears along the same line of sight. The other star is the pulsar (see Chapter 11) that remains from the supernova explosion. It spins 30 times per second, and one or the other of its two lighthouse beams sweeps across Earth every $\frac{1}{60}$ of a second, with the same frequency as your 60-cycle, alternating-current household electricity. (I think that's a "powerful" analogy.)

» **The North American Nebula,** NGC 7000, in Cygnus, the Swan

The North American Nebula (the name comes from its shape) is a faint but large H II region that you can see with the naked eye on a summer evening at a dark location with no moonlight. To glimpse it, use averted vision — look out of the corner of your eye.

» **The Northern Coal Sack,** in Cygnus, the Swan

The Northern Coal Sack is a dark nebula near Deneb, which is Alpha Cygni, the brightest star in Cygnus. You can recognize the Coal Sack by eye as a dark blotch against the brighter background of the Milky Way.

Don't skip the following nebulae located at moderate southern declinations. You can view them from many places in the Northern Hemisphere and from anywhere in the Southern Hemisphere:

» **The Lagoon Nebula,** Messier 8, in Sagittarius, the Archer

» **The Trifid Nebula,** Messier 20, in Sagittarius

The Lagoon Nebula and the Trifid Nebula are large, bright H II regions that you can see in the same field of view of your binoculars. Their best observation time is during summer evenings. A color photo shows that the Trifid has a bright red region and a separate, fainter blue region. The red area is the H II region, and the blue zone is a reflection nebula.

Great nebulae of the deep Southern Hemisphere include

» **The Tarantula Nebula,** in Dorado, the Goldfish

The Tarantula Nebula is in the Large Magellanic Cloud Galaxy, but it's such a huge and brilliant H II region that it's conspicuous to the naked eye for viewers in temperate southern and far southern latitudes. The Tarantula is another object to observe if you take a South Seas cruise, in addition to the Southern Cross and the Jewel Box star cluster (see the section "Star Clusters: Meeting Galactic Associates" earlier in this chapter).

» **The Carina Nebula,** in Carina, the Ship's Keel

The Carina Nebula, near the huge, unstable star Eta Carinae (see Chapter 11), is a large, bright H II region.

>> **The Coal Sack,** in Crux, the Cross

The Coal Sack, a dark nebula, is a large black patch, several degrees on a side, in the Milky Way. You can't miss it on a clear night with a dark sky, as long as you're deep in the Southern Hemisphere. (By the way, Crux, the Cross, is the official name for what most people call the Southern Cross.)

>> **The Eight-Burst Nebula,** NGC 3132, in Vela, the Sail

The Eight-Burst Nebula is a planetary nebula that's sometimes called the Southern Ring Nebula; it resembles the Ring Nebula in Lyra, mentioned earlier in this section, but is a bit fainter.

TIP

For some of the best color images of nebulae (as well as galaxies and deep space objects) ever photographed, check out the Hubble Heritage Image Gallery, at heritage.stsci.edu/gallery/gallery.html.

Another great source for photos of nebulae and other celestial objects is the Astronomy Picture of the Day website at apod.nasa.gov. My favorite Astronomy Picture of the Day is just right for this chapter (apod.nasa.gov/apod/ap170203.html). It's a night panorama of mountains and the Milky Way, with the Southern Cross, the planets Mars and Saturn, the Large and Small Magellanic Cloud galaxies, and more in view. If you look hard, you can just make out two observatory domes on the closest mountain. Can't recognize all these sky objects? Just place your cursor anywhere on the photo, and object names and constellation outlines appear. (I describe the Magellanic Clouds in the later section "Gawking at great galaxies.")

Getting a Grip on Galaxies

A large galaxy consists of thousands of star clusters and billions to trillions of individual stars, all held together by gravity. The Milky Way fits this bill as a large spiral system. But galaxies come in many shapes and sizes (see Figure 12-5 for sketches of several main types).

The main types of galaxies, based on shape and size, are as follows:

>> Spiral

>> Barred spiral

>> Lenticular

>> Elliptical

>> Irregular

>> Low surface brightness

I cover all these types in the following sections, as well as great galaxies to observe; the Milky Way's home, the Local Group; and even larger groups of galaxies, such as clusters and superclusters.

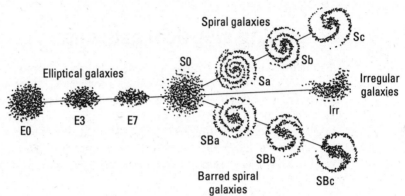

FIGURE 12-5:
Galaxies come in many different shapes and sizes.

Dinah L. Moché/Astronomy: A Self-Teaching Guide, Seventh Edition

Surveying spiral, barred spiral, and lenticular galaxies

Spiral galaxies are shaped like disks, with spiral arms winding through them. They may resemble the Milky Way, or their arms may be wound more or less tightly than our galaxy's spiral arms. The central bulge of stars in another spiral galaxy may be more prominent or less prominent, compared to the spiral arms. In Figure 12-5, spiral galaxies are labeled by their *Hubble types*, Sa, Sb, and Sc. (Yes, galaxy types are named after Edwin Hubble, too.) As you go from Sa to Sc in that sequence (or beyond, to Sd), the galaxies' spiral arms are less tightly wound and the central bulges are less prominent.

Spiral galaxies have plenty of interstellar gas, nebulae, OB associations, and open clusters, in addition to globular clusters. You can see a photo of spiral galaxies in the color section.

Barred spiral galaxies are spiral galaxies in which the spiral arms don't seem to emerge from the galaxy center; instead, they seem to protrude from the ends of a linear or football-shape cloud of stars that straddles the center. This star cloud is called the *bar.* Gas from outer parts of the galaxy may be funneled toward the center through the bar in a process that forms new stars that make the galaxy's central bulge, well, bulge even more. These galaxies include the types labeled as SBa, SBb, and SBc in Figure 12-5. The sequence from SBa to SBc (and beyond, to SBd, not shown in the figure) goes from barred spirals with tightly wound arms and relatively large central bulges to ones with open arms and small bulges.

Lenticular (lens-shape) galaxies are flattened systems with galactic disks, just like spiral galaxies. They contain gas and dust, but they don't have spiral arms. These galaxies are marked with their type, S0, in Figure 12-5.

Examining elliptical galaxies

Elliptical galaxies are shaped like a football, and that definition includes both the shapes of U.S. footballs and the shapes of soccer balls. Some ellipticals, in other words, are ellipsoidal in shape, roughly like a U.S. football, and some are spherical, like a soccer ball. Ellipticals can be beautiful sights, and I get a kick out of them. They contain many old stars and globular star clusters, but not much else. Elliptical galaxies are shown labeled with Hubble types from E0 to E7 in Figure 12-5. They form a sequence from the roundest ones at E0 to the flattest elliptical galaxies at E7.

Elliptical galaxies are systems in which star formation has largely or totally ceased. They have no H II regions, young star clusters, or OB associations. Imagine living in one of these dull galaxies, with nothing like the Orion Nebula to entertain you or give birth to new stars. And you probably won't find much on television, either.

The production of new stars may have ended in an elliptical galaxy because all the interstellar gas was used up in making the stars already in the galaxy. Or star formation may have ended because something blew out or swept out all the remaining gas suitable for making more stars. I say "suitable" because some elliptical galaxies, although they show no H II regions or groups of young stars, do have some extremely hot gas — so thin and hot, in fact, that it shines only in X-rays. Gas in this state doesn't readily condense into a star. And to tell the truth, some elliptical galaxies do display a number of bluish star clusters, which appear to be very young globular star clusters, much younger than any in the Milky Way.

TECHNICAL STUFF

One leading theory of elliptical galaxies — or at least of *some* elliptical galaxies — is that they form through the collision and merger of smaller galaxies. The collision of two spiral galaxies, for example, can produce a large elliptical galaxy, and shock waves from the event may compress large molecular clouds in the spirals, giving birth to huge clusters of hot, young stars — perhaps the very bluish star clusters found in some ellipticals. But the collision of a small spiral galaxy with a big spiral may lead only to the latter swallowing the former, making the spiral's central bulge even bigger.

A GALAXY IS A GALAXY IS A GALAXY

Writing *galaxy* and *galaxies* over and over gets repetitive. But what's a good synonym for *galaxy*? Some uninformed folks (or their editors) write *star cluster* to vary their prose, but that's wrong. And a large group of galaxies isn't a *galactic cluster*, an alternate term for an open cluster of stars inside a galaxy. Instead, astronomers call a large group of galaxies a *cluster of galaxies*, sometimes written as *galaxy cluster*. The cluster is composed of galaxies and is thus *galaxian*, but it isn't *galactic*.

As astronomers look out into space, we see many examples of colliding and merging galaxies. The farther out in space we look (so that we are examining the universe at more ancient times), the more prevalent the mergers seem to be. Apparently, galaxy collisions were common in the early universe and may have helped shape many of the galaxies we see today.

Looking at irregular, dwarf, and low surface brightness galaxies

Irregular galaxies have shapes that tend to be, well, strictly irregular. You may find the glimmerings of a little spiral structure in one of them, or you may not. They generally have plenty of cool interstellar gas, with new stars forming all the time. And they usually appear smaller than full-size spirals and ellipticals, with many fewer stars. You can see an irregular galaxy with its type, Irr, in Figure 12-5. The Large and Small Magellanic Clouds are two famous examples of irregular galaxies.

Dwarf galaxies are just what the name implies: itty bitty galaxies that may be mere thousands of light-years (or less) across. The types of dwarf galaxies include dwarf ellipticals, dwarf spheroidals, dwarf irregulars, and dwarf spirals. Snow White had only seven dwarfs, but the universe may have billions of dwarf galaxies. In our immediate neck of the woods, the Local Group of Galaxies, the most common galaxies are dwarf galaxies — just as in the Milky Way, the most common stars are the smallest stars, red dwarfs. The same probably holds true in the rest of the universe. And just as red dwarf stars are small, faint, and hard to see even when they're nearby, the same holds true for dwarf galaxies.

Dwarf galaxies are often unusually rich in dark matter, a mysterious substance (or substances) that I describe in Chapter 15. You don't see dwarf galaxies in Figure 12-5 because Edwin Hubble didn't include them when he made up the original diagram. He didn't include the next type that I list, the low surface brightness galaxies, because they hadn't been discovered yet. Hey, nobody's perfect.

Low surface brightness galaxies were recognized as a major variety in the 1990s. Some of them are as large as most other galaxies, yet they barely shine at all. Although they have a full tank of gas, they haven't produced many stars from the gas, so they don't appear as bright. Astronomers missed them for decades in surveys of the sky, but we're starting to pick them up now with advanced electronic cameras. Astronomers have found some very small low surface brightness galaxies, which are the least luminous galaxies of all. I call them "dim bulb galaxies." Who knows what else is out there that we haven't spotted yet?

Some astrophysicists think that much of the mass in the universe may be present in the form of low surface brightness galaxies that we haven't counted properly, like some groups of people who may be under-represented in the U.S. Census.

Gawking at great galaxies

TIP

To enjoy telescopic views of galaxies, use telescopes like those I recommend in the section "Enjoying Earth's best nebular views" earlier in this chapter. Big galaxies like the Andromeda Galaxy or the Triangulum Galaxy are great sights through a low f/number telescope (see Chapter 3). For viewing smaller galaxies, I recommend a telescope with a computer control feature that points the telescope to exactly the right place in the sky. *Sky & Telescope's Pocket Sky Atlas* and other atlases show where the bright galaxies are among the constellations.

The best galaxies for viewing from the Northern Hemisphere include the following. When I mention the season of the year for best viewing, I mean the season in the Northern Hemisphere. (Remember, it's fall in the Northern Hemisphere when Brazilians are enjoying spring.)

>> **The Andromeda Galaxy** (Messier 31; see Chapter 1), in Andromeda, a constellation named for an Ethiopian princess in Greek mythology

The Andromeda Galaxy is also called the Great Spiral Galaxy in Andromeda, and it was long known as the Great Spiral Nebula in Andromeda, or just the Andromeda Nebula. It looks like a fuzzy patch to the naked eye. You can see it in the autumn evening sky. From a dark observing site, you can trace it across about 3° — or about six times the width of the full Moon — on the sky with your binoculars. Don't try to view this galaxy during the full Moon — you will get a poor view, at best; wait until the Moon is just a crescent or is below the horizon. The darker the night, the more of the Andromeda Galaxy you can see.

>> **NGC 205 and Messier 32,** in Andromeda

NGC 205 and Messier 32 are small, elliptical companion galaxies of the Andromeda Galaxy. Some experts call them both dwarf elliptical galaxies, and some don't. (I wish they'd make up their minds.) M32 is spheroidal in shape, and NGC 205 is ellipsoidal.

>> **The Triangulum, or Pinwheel Galaxy** (Messier 33), in Triangulum, the Triangle

The Triangulum, or Pinwheel Galaxy, is another large, bright, nearby spiral galaxy, smaller and a little dimmer than the Andromeda Galaxy — and also a fine sight through binoculars in the fall.

>> **The Whirlpool Galaxy** (Messier 51), in Canes Venatici, the Hunting Dogs (see Figure 12-6)

The Whirlpool Galaxy is farther away and fainter than the Andromeda Galaxy and the Triangulum Galaxy, but you get a more glorious view of it through a high-quality small telescope. The Whirlpool is a *face-on* spiral, meaning that the galactic disk is pretty much at right angles to our line of sight from Earth; we look right down (or up) on it. With the larger telescopes at a star party (see Chapter 2), you should be able to make out its spiral structure from a distance of about 23 million light-years. Messier 51 is where the 3rd Earl of Rosse discovered the spiral structure of galaxies in 1845. (He just happened to own the largest telescope in the world.) Look for it on a fine, dark evening in the spring.

>> **The Sombrero Galaxy** (Messier 104), in Virgo, the Virgin

The Sombrero Galaxy looks like a bright, edge-on spiral galaxy, and that's how astronomers classified it until recently. However, now some astronomers think that the Sombrero is a giant elliptical galaxy that somehow has a structure resembling a spiral galaxy inside it. The "brim" of the Sombrero is the spiral structure's galactic disk. A dark stripe appears along the brim because the band of dark nebulae or "coal sacks" in the disk is edge-on to our line of sight. Look for the Sombrero in the spring, too; it's a bit farther away than the Whirlpool, but it still looks good through a telescope.

The following list presents the finest galaxies for observers in the Southern Hemisphere:

>> **The Large and Small Magellanic Clouds** (LMC and SMC) are irregular galaxies that orbit the Milky Way. The Large Magellanic Cloud is not only larger but also closer to Earth. It orbits a mere 163,000 light-years (give or take) from us. In fact, scientists believed for many years that the LMC was the closest galaxy to the Milky Way. (Today scientists know that three dim, miserable excuses for galaxies, the Sagittarius Dwarf Galaxy, the Canis Major Dwarf Galaxy, and the Ursa Major II Dwarf Galaxy, are even closer. However, we can barely discern those three objects in telescopic photos because the Milky Way is absorbing them.)

FIGURE 12-6:
The Whirlpool
Galaxy, photo-
graphed in
ultraviolet light by
the GALEX
satellite.

Courtesy of NASA/JPL/Caltech

The LMC and SMC look to the naked eye like clouds in the night sky. They are circumpolar in much of the Southern Hemisphere. In other words, at far southern latitudes, they never set below the horizon. If you go far enough south in South America or elsewhere in the Southern Hemisphere, the LMC and SMC are visible on every clear night of the year. Spot them by eye and then scan across them with binoculars and see how many star clusters and nebulae you recognize.

>> **The Sculptor Galaxy** (NGC 253) is a large, bright spiral galaxy and is one of the dustiest. Caroline Herschel, who also found eight comets, discovered it in 1783. Southern Hemisphere folks can look for it with binoculars or a telescope on a dark spring evening. Observers in the continental United States who have an unobstructed southern horizon can look for it low in the sky in autumn.

>> **Centaurus A** (NGC 5128) is a huge galaxy with a peculiar appearance: spheroidal, but with a thick band of dark dust across its middle. The galaxy is a powerful source of radio waves and X-rays and has been much studied with radio telescopes and with X-ray telescopes in orbiting satellites. Theorists have gone back and forth on whether it's an example of colliding galaxies. Some astronomers suspect that it may be another object like the Sombrero Galaxy: a giant elliptical galaxy with spiral structure within. Regardless of which theory

is correct, I think that Centaurus A has swallowed a smaller galaxy or two in its time, so watch from a safe distance. This object is most suitable for viewing during autumn in the Southern Hemisphere.

Discovering the Local Group of Galaxies

The Local Group of Galaxies, called the Local Group for short, has more than 50 members. It includes two large spirals (the Milky Way and the Andromeda Galaxy), a smaller spiral (the Triangulum Galaxy), their satellites (including the Large and Small Magellanic Clouds, as well as M32 and NGC 205), and lots of dwarf galaxies.

The Local Group isn't much as assemblages of galaxies go, but it's home, and it's the largest structure that we on Earth are gravitationally bound to (meaning that Earth isn't flying away from the Local Group as the universe expands). Just as the solar system isn't getting any bigger — because the Sun's gravity prevents the planets from moving outward or escaping — the Local Group holds strong because of the gravity of the three spiral galaxies and the smaller members. But all other groups and clusters of galaxies and distant individual galaxies *are* moving away from the Local Group at rates determined by a formula called *Hubble's Law* (named for You-Know-Who). Chapter 16 explains more about the away-ward movement.

The Local Group is about 3 megaparsecs across and centered near the Milky Way. A *parsec* is a length in space equal to 3.26 light-years, and *mega* means "million," so the Local Group is almost 10 million light-years wide. That dimension may sound large, but the Local Group is minuscule compared to the vast extent of the observable universe.

Clusters and superclusters of galaxies are much larger than the Local Group and can be photographed with a large telescope across billions of light-years in space. But the majority of all the galaxies in the universe, at least the readily visible ones, aren't in those big formations; they're located in small groups with only dozens of members or fewer, like the Local Group (which has about 50 or so at the latest count). So we appear to be in an average condition, as galaxy neighborhoods go.

Checking out clusters of galaxies

Most galaxies may be in small groups like the Local Group, but as astronomers survey the distant heavens with professional observatory telescopes, the formations that stand out are the clusters of galaxies. Most prominent are the so-called *rich clusters,* with hundreds and even thousands of galaxy members, each with its own complement of billions of stars.

The nearest large cluster of galaxies is the Virgo Cluster, spread out across the constellation of the same name and adjacent constellations. The cluster is about 54 million light-years away and contains more than 1,000 galaxies.

You can observe some of the biggest and brightest member galaxies of the Virgo Cluster with your own telescope. Messier 87 is one of the best sights: a spheroidal giant elliptical galaxy with a powerful jet of matter flying out at its center from the vicinity of a supermassive black hole. You can see M87 with amateur equipment, but not the jet at its center, unless you're a *very* advanced amateur. The galaxy appears to have swallowed some smaller ones, which may be why it's so big. Some galaxies like to start small and work their way up. Messier 49 and Messier 84 are two more Virgo Cluster giant ellipticals that you can observe, and Messier 100 is a large spiral galaxy in the cluster. Look for these galaxies on a dark evening in the spring in the Northern Hemisphere. Use a telescope with computer control that can home in on them for you. And if you don't trust computers, be sure you have a good star atlas that shows the galaxies.

Clusters of galaxies exist as far as our telescopes can see. At the limit of current technology in the early 21st century, we estimate that there are about 2 trillion galaxies in the observable universe, but nobody has counted them all — at least, nobody on our planet.

Sizing up superclusters, cosmic voids, and Great Walls

You may think that a large cluster of galaxies, up to 3 million light-years across, would be the max. But deep-sky surveys indicate that most or all galaxy clusters are grouped into larger forms, called *superclusters*. The superclusters aren't necessarily held together by gravity, but they haven't fallen apart, either. They appear to have long, filamentary shapes and flat, pancakelike shapes. A supercluster can contain a dozen clusters of galaxies, or hundreds of these clusters, and it can be 100 or 200 million light-years long.

We exist in the outer parts of the Local Supercluster, sometimes called the Virgo Supercluster, which is centered near the Virgo Cluster of Galaxies. The Virgo Supercluster is itself part of a larger structure called the Laniakea Supercluster.

The superclusters seem to be positioned on the edges of huge, relatively empty regions of the universe called *cosmic voids*. The nearest one, the Bootes Void, is more than 300 million light-years across. Many galaxies sit on its periphery, but we don't see very many inside the Void.

Astronomer Robert Kirshner discovered the Bootes Void. But when he was congratulated on the find, he reportedly said modestly, "It's nothing."

Some of the biggest superclusters, or groups of superclusters, are called *Great Walls.* The first Great Wall discovered is about 750 million light-years long (estimates differ). But other Great Walls, far out in the universe, may be larger. As far as astronomers know, the Great Walls don't display any Great Graffiti, but they have plenty to tell us about the origin of large structures in space and the early history of the universe. If only we understood the language.

Joining Galaxy Zoo for Fun and Science

When you know the basic types of galaxies, why not help astronomers sift through the numerous galaxy photos from modern sky surveys? As in other citizen science projects that I describe in Chapter 11, all you need is your brain, your personal computer, and access to the Internet.

By signing up for the Galaxy Zoo at www.galaxyzoo.org, you add your efforts to those of hundreds of thousands of people around the world who have helped or are helping professional astronomers study distant galaxies captured in images from telescopes on the ground and in space.

After enrolling in Galaxy Zoo, you study examples of galaxy images and learn how to classify them easily by comparison with sample sketches. Then you're set to examine other pictures and help astronomers uncover new facts about the universe. In Galaxy Zoo's first year, almost 150,000 volunteers, called *Zooites,* provided more than 50 million galaxy classifications, judging whether galaxies were spirals or ellipticals. Later, Zooite work overturned the earlier belief that nearly all elliptical galaxies are red. Their findings showed that some ellipticals are blue, which means that they contain a population of hot young stars.

One Galaxy Zoo volunteer, Dutch teacher Hanny Van Arkel, discovered an object in Leo Minor (the constellation of the Little Lion) that didn't look like any known type of galaxy. She has gone down in the history of astronomy as the discoverer of this strange object, called *Hanny's Voorwerp* (Dutch for "Hanny's Object"). Ms. Van Arkel has been lionized by astronomers. Now that's something to roar about!

As an added benefit if you become a Zooite, you get to peer at the galaxy pictures for free! You have to pay to visit most major planetariums, but there's never a charge at the Galaxy Zoo. (Don't pet the galaxies.)

» Getting the scoop on quasars

» Identifying different kinds of active galactic nuclei

Chapter **13**

Digging into Black Holes and Quasars

B lack holes and quasars are two of the most exciting and sometimes mystifying areas of modern astronomy. Lucky for us astronomers, the two subjects are related. In this chapter, I explain the connection between the two mysteries and provide you with information about active galactic nuclei, a group that quasars fall into.

You may never see a black hole through your own telescope, but I can guarantee that when you tell people you're an astronomer, they'll blurt out, "What's a black hole?" I mention black holes briefly in Chapter 11, but in this chapter, I offer you the whole treatment.

Black Holes: Keeping Your Distance

A *black hole* is an object in space whose gravity is so powerful that not even light itself can escape from within — which is why black holes are invisible.

You can fall into a black hole, but you can't fall out — you can't get out if you want to (and you *would* want to). You can't even call home, so E.T. (in the 1982 movie) was lucky he landed in California and not in a black hole.

Anything that enters a black hole needs more oomph than it can ever have to get back out. The formal name for that "oomph" is escape velocity. Rocket scientists use the term *escape velocity* to represent the speed at which a rocket or any other object must travel to escape Earth's gravity and pass into interplanetary space. Astronomers apply the term in a similar way to any object in the universe.

The escape velocity on Earth is 7 miles per second (11 kilometers per second). Objects with weaker gravity have slower escape velocities (the escape velocity on Mars is only 3 miles or 5 kilometers per second), and objects with more powerful gravity have higher escape velocities. On Jupiter, the escape velocity is 38 miles per second (61 kilometers per second). But the champion of the universe for escape velocity is a black hole. The gravity of a black hole is so strong that its escape velocity is greater than the speed of light (186,000 miles per second or 300,000 kilometers per second). Nothing, not even light, can escape from a black hole (because you need to travel faster than the speed of light to escape a black hole, and nothing — including light — travels faster).

In 2011, a team of physicists reported on an experiment in which they found indications that some neutrinos (a type of subatomic particle that I describe in Chapter 10) travel faster than the speed of light. Such a finding would contradict a certain amount of known physics if it were correct, but it wasn't. Experts later traced the error in the experiment to a loose electrical connector. The physicists didn't have a screw loose, but the situation was almost as bad.

Some scientists have proposed a hypothetical class of particles, called *tachyons*, that can move faster than light. In fact, if tachyons exist, they can never travel at *less* than light speed. But the theory that these particles exist isn't widely accepted, and none of them have been found.

Looking over the black hole roster

Scientists can detect black holes when they see gas swirling around them that's too hot for normal conditions, when jets of high-energy particles make their escape as though to avoid falling in (the jets come not from inside a black hole, but from the black hole's immediate surroundings), and when stars race around orbits at fantastic speeds, as though driven by the gravitational pull of an enormous unseen mass (which they are).

Scientists recognize three types of black holes (as I mention in Chapter 11):

>> **Stellar mass black holes:** These black holes have the mass of a large star (about three to a hundred times more massive than the Sun), and they result from the deaths of such stars.

>> **Supermassive black holes:** These whoppers range from hundreds of thousands to more than 20 *billion* times as massive as the Sun and exist at the centers of galaxies. They may have come from the mergers of many closely packed stars or the collapse of huge gas clouds when the galaxies formed. But nobody knows for sure.

>> **Intermediate mass black holes:** The few tentative detections of these objects fit with the idea that they range from about 100 to about 10,000 solar masses. Some of them may exist at the centers of globular star clusters (which I describe in Chapter 11). So far, they have not yet been convincingly detected.

Poking around the black hole interior

A black hole has three parts:

>> **The event horizon:** The perimeter of the black hole

>> **The singularity:** The heart of the black hole formed from the ultimate compression of all matter within it

>> **Falling objects:** Matter that falls from the event horizon toward the singularity

The following sections describe these parts in more detail.

The event horizon

The event horizon is a spherical boundary that defines the black hole. After an object enters the event horizon (see Figure 13-1), it can never get back out of the black hole or be seen by anyone on the outside.

The size of the event horizon is proportional to the mass of the black hole. Make the black hole twice as massive, and you make its event horizon twice as wide. If scientists had a way to squeeze Earth until it became a black hole (we don't, and if we did, I wouldn't tell how), our planet would have an event horizon less than ¾ of an inch (about 2 centimeters) across.

FIGURE 13-1:
One concept of a black hole, with arrows representing doomed matter falling in.

© John Wiley & Sons, Inc.

Table 13-1 offers a list of black hole sizes, in case you want to try some on. The two largest black holes in the table are at the centers of giant elliptical galaxies that are the brightest and most massive galaxies in the respective clusters of galaxies in which they're located. (I describe clusters of galaxies in Chapter 12.)

TABLE 13-1   Black Hole Measurements

Black Hole Mass in Solar Masses	Diameter of Black Hole in Miles	Diameter of Black Hole in Kilometers	Comment
3	11	18	Smallest stellar mass black hole
10	37	60	Typical stellar mass black hole
100	370	600	Largest stellar mass black hole
1,000	3,700	6,000	Intermediate mass black hole
4 million	15 million	24 million	Supermassive black hole at Milky Way center
6.3 billion	23 billion	37 billion	Supermassive black hole in M87 in the Virgo Cluster
21 billion	77 billion	120 billion	Supermassive black hole in NGC 4889 in the Coma Cluster

As far as scientists know, no black holes are smaller than about three solar masses and 11 miles wide.

Astronomers are planning to make the first image of an event horizon. They will use the Event Horizon Telescope (EHT) — actually a bunch of radio telescopes located across Earth but functioning in unison for the duration of the experiment. The EHT will be directed toward Sgr A*, the supermassive black hole in the center of the Milky Way. If it succeeds, the picture of the event horizon will test Einstein's Theory of General Relativity and provide new information on the black hole and perhaps on how its gravity is pulling in interstellar matter now. (The scientific term is *accreting*.)

The picture should look like a circular black area with a bright crescent on one side. The black circle is the dark "shadow" of the event horizon (we can't actually see the event horizon itself). The first try at making this observation occurred in April 2017, but the results weren't available as this book went to press. Not all complicated experiments come off as desired on the first attempt (or the second). Nevertheless, once the Milky Way's black hole has been studied, the likely next step will be to turn the same equipment on M87. This galaxy is a giant elliptical galaxy in the Virgo cluster, and has a much larger black hole but is located much farther from Earth.

The singularity and falling objects

Anything that falls inside the event horizon moves down toward the singularity. It merges into the singularity, which scientists believe is infinitely dense. We don't know what laws of physics apply at these immense densities, so we can't describe what the conditions are like. You might say we have a black hole in our knowledge.

Some mathematicians think that at the singularity there may be a *wormhole*, a passage from the black hole to another universe. The wormhole concept has inspired authors and movie directors to produce plenty of science fiction on the topic, but the writers and directors are just fishing. Most experts think wormholes don't exist. And even if they do, scientists have no way to see wormholes inside black holes or to inch our way down to them.

Surveying a black hole's surroundings

Here's what scientists observe in the vicinities of black holes:

1. **Gaseous matter falling toward the black hole swirls around in a flattened cloud called an *accretion disk*.**

2. **As the gas in the accretion disk gets closer to the black hole, it becomes denser and hotter.**

   The gas heats up because the gravity of the black hole compresses it, a process that occurs because friction increases as the gas gets denser. (The process resembles the way air conditioners and refrigerators work: When gas expands, it cools, and when it gets compressed, it gets hotter.)

3. **As the hotter and denser gas approaches the black hole, it glows brightly; in other words, the accretion disk shines.**

   Radiation from the accretion disk can take many forms, but the most common type for stellar mass black holes is X-rays. X-ray telescopes, such as the one onboard NASA's large Earth-orbiting satellite, the Chandra X-ray Observatory, detect X-rays and allow scientists to pinpoint black holes. You can see the X-ray images from Chandra at `http://chandra.harvard.edu`, the website of the Chandra X-ray Center, by clicking on the Photo Album link.

So although you can't actually see a black hole through a telescope, you can detect radiation from the accretion disk of hot gas swirling around it. For stellar mass black holes, however, you need to have an X-ray telescope that's up in space. X-rays don't penetrate Earth's atmosphere, so the telescope needs to be above the atmosphere. But the accretion disks around supermassive black holes in some nearby galaxies emit much of their light in the ultraviolet and the optical, so they can be observed with moderate-sized telescopes here on the ground. To the casual observer, it will simply appear as though a bright star has taken up residence in the center of the galaxy.

TECHNICAL STUFF

Bare black holes may exist in space, with no gas swirling into them. If so, astronomers can't see them unless they just happen to pass in front of a background star or galaxy under observation. In that case, you can infer that the black hole exists because you see the effect of its gravity on the appearance of the background object. (You may see the background object get briefly brighter, for example, as I describe in Chapter 11 when I discuss gravitational microlensing.) But this situation is a rare coincidence. Another rare occurrence involves an unsuspecting star wandering too close to a black hole and being ripped to shreds, creating a temporary accretion disk. These events are called *tidal disruptions* and are another way for invisible black holes to show themselves. But don't hold your breath waiting for a black hole to bare itself or light up for your inspection. (For more on tidal disruption events, flip to the "Watching stars get swallowed by black holes" section later in this chapter.)

Warping space and time

You can think of a black hole as a place where the fabric of space and time is warped. A *straight line* — which is defined in physics as the path taken by light moving through a vacuum — becomes curved in the vicinity of a black hole. And as an object approaches a black hole, time itself behaves oddly, at least as perceived by an observer at a safe distance.

Suppose that, as you move at a safe distance, you launch a probe toward a black hole from your spaceship. A big electric billboard on the side of the probe displays the time given by a clock in the probe.

From the spaceship, you watch the clock through a telescope as the probe falls toward the black hole. You see that the clock runs slower and slower as the probe approaches the black hole. In fact, you never actually see the probe fall in. You see it get redder and redder as the light from the electric billboard is redshifted by the powerful gravity of the black hole — not because of the Doppler effect (which I describe in Chapter 11) but because of a phenomenon called *gravitational redshift*. The light of the billboard shifts toward longer wavelengths, just as the Doppler effect makes the light from a star moving away from the observer seem to shift toward longer wavelengths. After a while, gravity redshifts the glow of the electric billboard to infrared light, which your eyes can't detect.

Now consider what you'd see if you traveled onboard the falling probe. (Don't try this at home — in fact, don't try it anywhere.) You can watch the clock inside the probe, and you can peek back along your path through a window. You, the ill-fated onboard observer, see that the clock runs normally. You don't perceive that it runs slow at all. As you look out the window at the mother ship and the stars, they all seem to be blueshifted. You're blue at the thought that you can never go home again. You pass through an invisible boundary (the event horizon) around the black hole in no time.

A person on the mother ship never sees you enter the black hole; you just appear to get closer and closer to the hole. But on the falling probe, you can tell that you've dropped right in. At least, you can if you're still alive. Ultimately, *tidal force*, an effect of the immense gravity, pulls apart anything that falls into a black hole; at least, along one dimension (the direction toward the singularity), you're pulled apart. To make matters worse, in the other two spatial dimensions, tidal force squeezes you together unmercifully.

If you enter the black hole feet first, tidal force stretches you out (if you're not already pulled apart) until you're more than tall enough to be drafted as a center in the National Basketball Association. But from bellybutton to back and from hip to hip, you get squeezed together like coal turning into diamond under immense pressure inside Earth, only worse. This experience is no gem.

Small or stellar mass black holes are the most deadly variety, just as some small spiders are more poisonous than big tarantulas. If you fall toward a stellar mass black hole, you're torn apart and squeezed together before you enter — astronomers like to call this "spaghettification" — and you never get to see the universe disappear before you die. But falling into a supermassive black hole is a different experience. You get to fall inside the event horizon and see the universe black out before you suffer the tidal fate (or is it the fatal tide?).

Considering that black holes are all around us in the universe and that they have such fascinating and strange properties, you can see why scientists want to study them, but from a safe distance.

Detecting black hole collisions

When two black holes orbit their shared center of mass, they emit gravitational radiation, causing them to spiral ever closer together. The gravitational radiation, predicted by Einstein, is like the ripples in a pond when you throw a stone in. However, these ripples aren't waves in water but a spreading disturbance in space-time itself. As the black holes spiral toward the center, they orbit ever faster and finally merge, forming a single, bigger black hole.

The mass of the black hole produced by two smaller black holes that come together is a little less than the sum of the masses of the two original black holes. In the first known black hole collision, which was detected on September 14, 2015, the individual masses were 36 and 29 times the mass of the Sun, respectively, which add up to 65 solar masses. But the larger black hole that they became has only 62 solar masses. A mass equal to three times that of the Sun was released in the form of gravitational waves during the final collision, which spread out through the universe at the speed of light, reaching Earth in 2015, 1.3 billion years after the collision occurred. Inside the Sun, 5 million tons of hydrogen are converted into energy every second, as Chapter 10 describes. The black hole collision was far more powerful: Three solar masses is equivalent to 6 billion billion billion tons!

The observation of the black hole collision was no fluke. A few months later, another one was detected: the merger of black holes with 14 and 8 solar masses, respectively. The final black hole amounted to 21 solar masses, so one solar mass was lost in the form of gravitational waves.

The facility that made these first direct detections of gravitational waves is the Laser Interferometer Gravitational-Wave Observatory (LIGO), which consists of two large laboratories in Louisiana and Washington State that work together to make a detection, like the scattered radio telescopes in the Event Horizon Telescope, which I describe earlier in the chapter. You can't make detections like this with home equipment: To observe gravitational waves, LIGO has to measure changes in a 2½-mile length that amount to only one ten-thousandth of the diameter of a proton!

REMEMBER

LIGO can make even better measurements if you help improve its performance. You just need to sign up for the Gravity Spy citizen science project. Go to `www.zooniverse.org/projects/zooniverse/gravity-spy` and click on "Learn more"; if you get the hang of it, you can pull your own weight in the study of gravity.

Watching stars get swallowed by black holes

Sometimes when a star comes too close to the supermassive black hole at the center of a galaxy, tidal force of the black hole tears the star apart. (I explain tidal force in the earlier section "Warping space and time.") In that case, astronomers can observe a *tidal disruption event* (TDE), meaning a bright flare in visible, ultraviolet light, X-rays and/or radio waves, that may last for months. When a TDE occurs, about half of the star falls into the black hole; the other half is thrown out into the galaxy around the black hole.

The distance from the center of a black hole within which a star will be torn apart is called the *tidal disruption radius* because as the star approaches, that's the distance at which the tidal force is strong enough to do the job. But a black hole also has a radius within which we can't see something falling in. That's the radius of the event horizon, which I define earlier in this chapter.

In the biggest black holes, the event horizon is outside the tidal disruption radius, so a star coming too close to the black hole vanishes from view before it can be disrupted — we don't see a TDE. But for most black holes, the tidal disruption radius extends outside the event horizon, so when a star falls in we see the great flare.

The exact details depend on what kind of a star is falling in or being disrupted and on how fast the black hole is spinning, if at all. But here are two simple examples:

» A star like the Sun would be disrupted if came close enough to the 4-million solar mass black hole in the center of the Milky Way. But it would have to get to less than one AU (astronomical unit; about 93 million miles) from the hole.

» A similar Sun-like star can never get close enough to a black hole of 100 million times the solar mass or more to produce a TDE that we can see. The star would vanish inside the event horizon before being disrupted.

These examples apply when the black holes don't spin. For a spinning black hole, the disruption can be even more spectacular than usual, and for the most massive black holes, the spin of the black hole may cause the disruption to occur earlier, when the star is still outside the event horizon and thus visible.

More than 20 actual or possible TDEs have been observed, and the number found is increasing as astronomers have learned the best ways to look for them.

Quasars: Defying Definitions

Scientists have at least two definitions for quasars:

>> **The original definition:** *Quasar* is an abbreviation or acronym for "quasistellar radio source" and means a celestial object that emits strong radio waves but looks like a star through an ordinary visible light telescope (see Figure 13-2).

The original definition of *quasar* has become outdated because, at most, 10 percent of all objects that we now call quasars fit this definition. The other 90 percent don't emit strong radio waves. Astronomers call them radio-quiet quasars.

>> **The current definition:** A *quasar* is an *active galactic nucleus,* meaning a supermassive black hole with an accretion disk that is fed by inflowing matter from the surrounding galaxy.

FIGURE 13-2:
A quasar in Sculptor shines at the center of this image.

Courtesy of ESO, Digitized Sky Survey 2 and S. Cantalupo (UCSC)

The current definition of *quasar* reflects the fact that, after decades of puzzling over quasars, astronomers have concluded that they're associated with giant black holes at the centers of galaxies. Matter that falls into the black hole releases enormous energy, and the observed energy sources are what astronomers call quasars.

The flow of matter into the black hole in a quasar can vary. When the flow is ample, the quasar shines with as much as 10 trillion times the energy the Sun releases each second. When the flow is low, however, the quasar fades away until it's revived with a good meal at a later time.

Measuring the size of a quasar

All quasars produce strong X-rays; about 10 percent produce strong radio waves; and they all emit ultraviolet, visible, and infrared light. All the emissions can vary in strength over years, months, and weeks, and even over times as short as hours or days.

The fact that quasars often change significantly in brightness over the course of a single day indicates to scientists something of surpassing importance: A quasar must be no larger than about one *light-day,* or the distance that light travels through a vacuum over the length of a day. And a light-day is only 16 billion miles (26 billion kilometers) long, which means that a quasar, producing as much light as 10 trillion suns, or 100 times as much light as the Milky Way Galaxy, isn't much bigger than our solar system, which is a tiny part of the Galaxy.

A quasar much bigger than a light-day can't fluctuate markedly in such a short time, any more than an elephant can flap its ears as rapidly as a hummingbird flutters its wings.

Getting up to speed on jets

Quasars that are strong radio sources often display *jets,* or long, narrow beams in which energy shoots out of the quasars in the form of high-speed electrons and perhaps other rapidly moving matter. Often the jets are lumpy, with blobs of matter moving outward along the beams. Sometimes the blobs appear to move at more than the speed of light. This *superluminal motion* is an illusion related to the fact that the jets, in such cases, are pointing almost exactly at Earth; the matter in them actually moves close to light speed, but not faster than light.

TIP

You can browse the best images of jets from quasars as detected by radio telescopes in the Image Gallery of the National Radio Astronomy Observatory, at images.nrao.edu.

Exploring quasar spectra

Many books say that a quasar has very broad lines in its spectrum, corresponding to redshifts and blueshifts of gas in turbulent motion within the quasar at up to 6,000 miles per second (10,000 kilometers per second). This statement isn't always true. Quasars come in a variety of types, and some don't have broad spectral lines. (See Chapter 11 for more on spectral lines.)

The broad spectral lines, however, are an important trait of many quasars and are a clue to their relationship to other objects, as I describe in the next section.

Active Galactic Nuclei: Welcome to the Quasar Family

For years after the discovery of quasars, astronomers argued about whether they're located in galaxies or are separate from galaxies. Today we know that quasars are indeed always located in galaxies because technology has improved to the point that we can make a telescopic image that shows both a quasar and the galaxy around it. The latter is called the quasar's *host galaxy*. (Previously, because a quasar can be 100 times brighter than its host galaxy — or even brighter — hosts got lost in the glare of their quasar guests in telescopic photos.)

TECHNICAL STUFF

Electronic cameras, which can record a greater range of brightnesses in a single exposure than photographic film, made the discovery possible.

Quasars are an extreme form of what astronomers now call *active galactic nuclei* (*AGN*). The term designates the central object of a galaxy when the object has quasarlike properties, such as a very bright starlike appearance, very broad spectral lines, and detectable brightness changes.

Sifting through different types of AGN

Scientists use the following main terms to describe AGN:

>> **Radio-loud quasars ("original quasars") and radio-quiet quasars (90 percent or more of quasars):** These two types are similar kinds of objects, with and without strong radio emission. They're often located in spiral galaxies, although some can be found in ellipticals (see Chapter 12 for more on galaxies). No quasar is visible in the Milky Way Galaxy, but we have detected a 4-million-solar-mass

black hole at its center, called Sagittarius A*. I list that supermassive black hole in Table 13-1.

» **Quasistellar objects (QSOs):** Some astronomers lump together the radio-loud and radio-quiet quasars as QSOs.

» **OVVs:** *Optically violently variable* quasars are quasars with jets that happen to point directly at Earth. These quasars undergo even more pronounced rapid brightness changes than ordinary, garden-variety quasars. Think of firefighters struggling to direct a firehose at a person whose clothes are on fire. The water pressure may be unstable, with the water pulsing a bit. The stream from the hose may look pretty steady to spectators from the side, but the person on the receiving end feels every fluctuation in the flow as the oncoming water batters him. OVVs are the firehoses of quasardom — they make the biggest splash.

» **BL Lacs:** Astronomy lingo for *BL Lacertae objects,* BL Lacs, as a group, are AGN that resemble BL Lacertae. BL Lacertae changes in brightness; for years, scientists thought it was just another variable star in the constellation Lacerta (it looks like a star in photographs of the sky). They later identified it as a strong source of radio waves and eventually determined that BL Lacertae was the active nucleus of a host galaxy that had been lost in its glare until improved technology made it possible to photograph the galaxy.

Unlike most quasars, a BL Lac doesn't have broad spectral lines. And its radio waves are more highly polarized than those from ordinary radio-loud quasars (except the OVVs, which may be just extreme cases of BL Lacs) — *polarized* means that the waves have a tendency to vibrate in a preferred direction as they travel through space. Unpolarized waves vibrate equally in all directions as they move. At the ballpark, you can't tell the players without a scorecard, as the saying goes; at the observatory, you have to check polarization to know your radio-loud quasars from your BL Lacs.

» **Blazars:** This term covers both OVVs and BL Lacs. The two quasar types have many similarities. Both are highly variable in brightness, their jets point directly at Earth, and they're both radio-loud.

Unlike other AGN, blazars shine very bright and sometimes brightest in gamma rays. Although blazars are fairly rare, they nevertheless make up the majority of known gamma ray sources beyond the Milky Way. (Gamma rays are the highest-energy form of light, more energetic even than X-rays.)

Do we really need a term to combine OVVs and BL Lacs? I'm not so sure. My friend Dr. Hong-Yee Chiu became famous among scientists for coining *quasar.* His friend Professor Edward Spiegel coined *blazar* a few years later. If you discover a new kind of object or write one of the leading studies on it, you may get to name it, too. Adding *ar* to your name isn't allowed; the term should be descriptive of the scientific properties of the object, not the astronomer.

>> **Radio galaxies:** These galaxies have relatively dim active galactic nuclei that nevertheless produce strong radio emissions. Most of the strongest radio-emitting galaxies are giant elliptical galaxies. Often they have beams or jets that transport energy from the AGN to huge lobes of radio emission, empty of stars, far outside and far larger than the host galaxy itself. Usually there is a lobe on one side of the galaxy and a second lobe on the opposite side.

>> **Seyfert galaxies:** These spiral galaxies have an AGN at their centers. A Seyfert AGN is like a quasar, with broad spectral lines and rapid brightness changes. It may be as bright as the host galaxy, but not 100 times brighter like a quasar, so the host isn't lost in the Seyfert nucleus's glare.

Examining the power behind AGN

All the different types of AGN have one thing in common: They're powered by energy that's generated in the vicinity of the supermassive black holes at their centers.

Near the supermassive black hole, stars and gas orbit the center of the host galaxy at immense speeds, a circumstance that enables astronomers to measure the black hole's mass. With telescopes such as the Hubble, astronomers can determine the velocities of the orbiting stars or orbiting gas by measuring Doppler shifts of the light from the stars or the gas (see Chapter 11 for more about the Doppler effect). The speeds indicate the mass of the central object. Stars at the same distance from the center of a less-massive black hole orbit at a slower pace.

In a quasar or a radio galaxy of the giant elliptical type, the supermassive black hole often attains a billion or more solar masses. In Seyfert galaxies, the black hole mass is often around 1-10 million solar masses.

The black hole makes it possible for the AGN to shine, but only the matter falling into the black hole actually powers the glow. It may take matter with ten times the mass of the Sun falling into the black hole each year to make a quasar shine.

If no material falls into the black hole, there is no AGN to reveal itself by producing a bright glow, radio emission, high-speed jets, or strong X-rays. Similar to kids who depend on their school lunches for the energy to perform in class, the black holes shine only when matter falls into them at a sufficient rate. Supermassive black holes may be lurking at the centers of most galaxies, but in most cases, matter isn't feeding them, so astronomers see quasars or other kinds of AGN in only a small fraction of galaxies.

Proposing the Unified Model of AGN

The *Unified Model of Active Galactic Nuclei* is a theory proposing that many kinds of AGN are, in fact, the same kind of object, which looks different when viewed from different angles. According to the Unified Model, when we look at AGN from different directions with respect to their accretion disks and their jets, they look different, just as a man you see face-on looks different than the same guy in profile. Everyone has a good side. The theory also proposes that black holes are sucking in matter at different rates, so some AGN (which are getting more matter per second than others) are brighter than the others for that reason alone. Dozens of astronomers write papers on the Unified Model every year; some find evidence for the theory, and some find evidence against it.

I think evidence points to real differences among the different types of AGN, but I also think they have many basic similarities. Astronomers need more information before we can unite around the Unified Model or any other theory of AGN. In the meantime, what do you think? Your taxes pay for much of this research, which is underway in just about every developed nation, so you're entitled to an opinion.

WHAT CAME FIRST: THE BLACK HOLE OR THE GALAXY?

An important discovery brightened the day of every quasar fan. Experts found a simple mathematical relationship between a supermassive black hole and the galaxy that surrounds it. The central region of most galaxies is called the *bulge*. Even a relatively flat spiral galaxy may have a central bulge, which can be big, middling, or small. An elliptical galaxy is considered all bulge. Astronomers have found that the mass of a black hole at the center of a bulge is usually close to one-fifth of 1 percent of the mass of the whole bulge. It seems as though every galaxy has to pay a tax of 0.2 percent to its black hole. (I wish the Internal Revenue Service let me off that cheaply.) An exception to this tax: Some spiral galaxies with small or nonexistent bulges don't have detectable central black holes. One good example is the Triangulum or Pinwheel Galaxy, which I describe in Chapter 12. And there are some "overmassive" black holes with masses higher than the 0.2 percent rule.

The unexpected relationship between black hole masses and galaxy bulge masses must be related to how they form, but astronomers aren't sure how. Does a big galaxy form around a big black hole? Or do big black holes form inside big, bulgy galaxies? Astronomers are arguing over this now in a dispute that I call the "battle of the bulges." Very likely, the black holes may grow larger as the bulges themselves grow when galaxies are young.

4

Pondering the Remarkable Universe

Find out what astronomers are doing in their Search for Extraterrestrial Intelligence (SETI) and how they are studying exoplanets that orbit stars beyond the Sun.

Explore the mysterious topic of dark matter and antimatter.

Consider the entire universe — its beginnings, its shape, and its future.

» Exploring (and participating in) SETI projects

» Hunting for extrasolar planets and seeing what they're like

Chapter **14**

Is Anybody Out There? SETI and Planets of Other Suns

The universe is both vast and varied. But do we share these starry realms with other thinking beings? Anyone who tunes in to *Star Trek* or frequents the local multiplex already knows Hollywood's answer: The cosmos is cluttered with aliens (many of whom seem to speak excellent English).

But what do scientists say? Are there really extraterrestrial beings out there? Many researchers believe that the answer is "yes," and some are looking for evidence. Their quest is known as SETI (rhymes with "Betty"), the Search for Extraterrestrial Intelligence. Other scientists are searching for traces of primitive life on Mars. Some experts think that the moons Europa and Enceladus (of Jupiter and Saturn, respectively) may harbor microbial life as well, but SETI seeks advanced civilizations capable of broadcasting into space. (For more about Europa and Enceladus, see Chapter 8; I cover Mars in Chapter 6.)

Why are many scientists optimistic about the possible existence of aliens? The upbeat attitude derives from the fact that our place in the cosmos is thoroughly unremarkable. The Sun may be an important star to us, but it serves as a bit player in the universe. The Milky Way galaxy hosts billions of similar stars. If this number fails to impress you, note that hundreds of billions of *other* galaxies are within range of our telescopes. The bottom line is that far more Sun-like stars are sprinkled through the visible universe than we have blades of grass on Earth. To assume that our blade of grass is the only place where something interesting happens is simply naive. Distressing as it may be to our self-esteem, Earth may not be the intellectual nexus of the universe.

How can earthlings find our brainy brethren? You can't go visit their likely homes. Rocketing off to distant star systems, although a work-a-day staple of science fiction, is actually quite difficult. The speed of earthly rockets, an impressive 30,000 miles per hour, is less impressive when you reckon that it would take these craft 1,000 centuries to reach Alpha Centauri, the nearest stellar stop on the tour of the universe. Faster rockets take less time, but they consume more energy — a lot more energy. (I explain a possible solution in the later section "Catching Proxima fever: Focusing on red dwarfs.")

Almost 60 years have passed since astronomer Frank Drake made the first significant efforts to put us in touch with aliens. So far, we haven't heard a single confirmed extraterrestrial peep. But the search has been limited. As technology advances, the chance for success increases. Someday soon, astronomers may puzzle over a signal received from the cold depths of space. Maybe it will be a message about the meaning of life or a complete summary of the laws of physics, including some as yet unknown to us. But one thing is for sure: The signal will show us that we're not the only kids on the galactic block.

Using Drake's Equation to Discuss SETI

Although we can't visit distant civilizations, astronomers seek evidence of technically sophisticated space aliens by listening for their radio traffic. In 1960, Frank Drake attempted to listen in on cosmic communications by using an 85-foot (in diameter) radio telescope in West Virginia. If you've seen the movie *Contact,* you know that a radio telescope may resemble a seriously bulked-up backyard satellite dish (see Figure 14-1), although some have other shapes. Drake connected his antenna to a new, sensitive receiver working at 1,420 MHz (located in what's called the microwave region of the radio spectrum) and pointed the telescope at two nearby Sun-like stars in what he called Project Ozma.

FIGURE 14-1:
With the right kind of receiver on a radio telescope, astronomers can listen for signals from other societies.

Courtesy of Seth Shostak

Drake didn't hear any aliens during Project Ozma, but he provoked a great deal of enthusiasm within the scientific community. A year later, in 1961, the first major conference on SETI was held. Drake tried to organize the meeting by distilling all the unknowns of the search into a single equation, now known as the *Drake Equation*. (For the mathematically inclined, I provide this simple little formula in the sidebar "Diving into the Drake Equation" later in this chapter.) Its logic is easy. The idea is to estimate N, the number of civilizations in our galaxy that use the radio airwaves now. N clearly depends on the number of suitable stars in the galaxy, multiplied by the fraction that have planets, multiplied by the number of . . . well, you can read about it in the sidebar.

Drake's Equation seems quite straightforward. But although scientists may know or can safely guess at the values of the first few terms (such as the rate at which stars capable of hosting planets form and the fraction of such stars that actually have planets), we don't have any real knowledge of such details as the fraction of life-bearing planets that can develop intelligent life or the lifetime of technological societies. So we don't know the "answer" to Drake's Equation, but it's a great way to frame your thinking about civilizations in our galaxy.

DIVING INTO THE DRAKE EQUATION

Scientists often use Frank Drake's nifty little formula as the basis for discussions about SETI and the chances that humans will ever make contact with extraterrestrial intelligent life. The equation is quite simple and doesn't require any math beyond what you mastered in the eighth grade.

The equation computes *N*, the number of civilizations active in the Milky Way galaxy that are broadcasting loud enough for us to detect. As with Shakespeare's name, which the playwright spelled three different ways in just one document, several versions of Drake's Equation exist, but here's a reasonable formulation:

$$N = R^* f_p n_e f_l f_i f_c L$$

- *R** is the rate at which long-lived stars, suitable for hosting habitable planets, form in our galaxy. Some experts think that this number is one or two stars formed per year in the Milky Way, but one study came up with a value of seven.

- *fp* is the fraction of stars (usually expressed as a percentage) that *have* planets. Astronomers don't know what this number is, but estimates based on recent planet searches indicate that the fraction is over 50 percent and may even approach 100 percent.

- *ne* is the number of planets per solar system that can incubate life. In our solar system, the number is at least one (Earth) and could be more if you count Mars and some of the moons of Jupiter and Saturn. But in another system, who knows? A typical guess is one, but recently astronomers found a star with seven exoplanets, three of which could be habitable. So maybe *ne* is larger than one.

- *fl* is the fraction of habitable planets that actually develop life. We can reasonably assume that many of them do, although some astronomers think that life is actually quite rare.

- *fi* is the fraction of planets with life that develop intelligent life. This number is controversial, of course, because intelligence may be a rare accident in biological evolution.

- *fc* gives the fraction of intelligent societies that invent technology (in particular, radio transmitters or lasers). Probably many of them do.

- *L*, the final term, is the lifetime of societies that use technology. This term is a matter of sociology, not astronomy, of course, so your guess is as good as the author's — maybe better.

The number *N* depends on your choice of values for the various terms. Pessimists think that *N* may be only 1 (meaning we're alone in the Milky Way galaxy). Carl Sagan figured it to be a few million. And what does Drake himself say? At one time, his guess was

SETI Projects: Listening for E.T.

Most of the modern SETI efforts follow in Frank Drake's footsteps. In other words, they use large radio telescopes to try eavesdropping on alien civilizations.

Why use radio? Radio waves move at the speed of light and easily penetrate clouds of gas and dust that fill the space between the stars. In addition, radio astronomy receivers are very sensitive. The amount of energy required to send a detectable signal from star to star (assuming that aliens use a transmitting antenna at least a few hundred feet in size) is no more than what your local television station pumps out.

Note: SETI planning has often stressed the idea of targeting Sun-like stars, but recently Earth-like planets have been found orbiting red dwarf stars. SETI astronomers are now targeting thousands of red dwarfs, as I describe later in this section. (You can read about red dwarf stars in Chapter 11.)

Assuming that researchers do get an interstellar ping, how do they recognize it? They don't expect to receive the value of *pi* or some other simple message that proves that the aliens completed junior high. SETI researchers mostly look for narrow-band signals.

A narrow-band signal occurs at one narrow spot on the radio dial, just like your favorite AM or FM station. Only a radio transmitter makes narrow-band emissions. Quasars, pulsars, and other natural radio sources in space all emit radio waves, but their natural static is spread out in frequency — splattered all over the radio spectrum. Narrow-band signals are the mark of transmitters. And transmitters are the mark of intelligence. It takes intelligence (not to mention a soldering iron) to build a transmitter.

Another criterion that SETI researchers insist on before they can claim to have received a true alien broadcast is that the signal be persistent. In other words, every time they point their telescopes at the source of the signal, they find it. If their meters register only once, the signal is impossible to confirm. They may count it as interference from telecommunication satellites, a software bug, or a prank. In the following sections, I discuss several SETI projects and explain how you can help with the search.

The flight of Project Phoenix

Project Phoenix, which operated from 1995 to 2004, used several radio telescopes, including the 1,000-foot dish at Arecibo, Puerto Rico (Figure 14-2), to make a targeted search for radio emissions from about 750 stars. A *targeted search* maximizes the time spent listening in what you think are the most likely directions to detectable civilizations. But you may guess wrong, so other searches scan as much of the sky as possible in the hope that the one signal you may detect comes from a place you'd never think of and that the signal is so strong you don't have to listen in its direction for more than a brief time. We'll never know which kind of search is better until one of them succeeds.

FIGURE 14-2:
A view of the massive Arecibo radio telescope in Puerto Rico that participated in *Project Phoenix*.

Courtesy of Seth Shostak

Phoenix and other SETI experiments operated in the microwave radio frequency range, but they didn't warm any leftovers. The microwave range is the preferred hunting ground for most SETI scientists because

>> The universe is rather quiet at microwave frequencies; you encounter less natural static, a fact that E.T. should also know.

>> A natural signal generated by hydrogen gas occurs at 1,420 MHz, a microwave frequency. Because hydrogen is by far the most abundant element in the cosmos, every alien radio astronomer should be aware of this natural marker — and may be tempted to get our attention (or the attention of other civilizations in space) by sending out a signal near its frequency on the dial.

But facing facts, scientists really don't know *exactly* where the extraterrestrials may tune their transmitters. To cover as much of the dial as possible, *Project Phoenix* checked out many millions of channels at once.

In the end, *Phoenix* found no persistent, clearly extraterrestrial signal. But this effort taught researchers how to build instruments for more powerful searches, such as the Allen Telescope Array (see the next section).

Space scanning with other SETI projects

Today several SETI programs dot the astronomy landscape:

>> The *Breakthrough Listen* project is a ten-year effort, begun in 2015, to search for extraterrestrial radio communications with the 328-foot Green Bank Telescope in West Virginia and the 210-foot Parkes Radio Telescope in Australia (both of which I describe in Chapter 2.) The search eventually will cover over one million stars in the Milky Way, with special attention to nearby stars. It also targets the central regions of nearby galaxies, although the prospects of detecting civilizations beyond the Milky Way seem very slim. You can find out what *Breakthrough Listen* is up to at any time by heading to the Berkeley SETI Research Center website (seti.berkeley.edu). Then click on "Breakthrough Listen" and scroll down to "Check the status of our main telescopes."

>> The *Automated Planet Finder (APF)* is a 94-inch robotic telescope at Lick Observatory on Mount Hamilton in California. As the name implies, it was built to discover exoplanets. (I explain how that's done in the later section "Finding exoplanets.") However, *Breakthrough Listen* is also using the APF to look for laser signals from extraterrestrial civilizations. Maybe aliens prefer laser communications to old-fashioned radio. APF may shed some light on that question.

>> The *SETI Institute* in Mountain View, California, is observing 20,000 red dwarf stars to catch radio signals that may come from planets orbiting the tiny, cool stars. (I describe red dwarfs in Chapter 11 and explain their new importance in the search for extraterrestrial life in the section "Discovering Alien Worlds" later in this chapter.) The work is underway with the Allen Telescope Array (ATA), a set of 42 dish antennas (each about 20 feet) wide, at Hat Creek Observatory in northern California. You can check the ATA's current or most recent observation by heading to the SETI Signal Searching page at setiquest.info. There you learn what the array is listening to, view pictures of the ATA from web-cams, and get info from the observer about whether the observing conditions are good or whether things aren't working right (for example, there may be radio interference).

>> *Astropulse*, operated by the University of California, Berkeley, is searching for alien signals that may take the form of extremely fast pulses of radio waves, perhaps one microsecond (millionth of a second) long. Natural radio emissions from pulsars or other celestial objects may also turn up in this search, and the astronomers have to learn to distinguish natural pulses from those an extraterrestrial civilization may artificially generate. This project uses the 1,000-foot Arecibo radio telescope in Puerto Rico (see Figure 14-2).

>> The *Search for Extraterrestrial Radio Emissions from Nearby Developed Intelligent Populations* (nicknamed *SERENDIP*), run by the University of California, Berkeley, uses the Arecibo and Green Bank Telescopes in "piggyback" mode, collecting data from whatever directions the telescopes happen to point. Thus, *SERENDIP* researchers can use the telescopes for SETI even when other astronomers direct the telescopes at pulsars, quasars, or other objects that aren't on a SETI target list. Of course, the loudest broadcasting aliens may be on a planet that's only or best observable from the Southern Hemisphere. So there's also *Southern SERENDIP*, run by the SETI Australia Centre at the University of Western Sydney. It piggybacks on the Parkes Radio Telescope.

>> *FAST*, the *Five-hundred-meter Aperture Spherical radio Telescope* in Guizhou Province, China, should be the next important contributor to SETI. It resembles the Arecibo radio telescope, but is much larger (1,640 feet). According to David Dickenson at www.skyandtelescope.com, "a Nimitz-class aircraft carrier could easily float in the 500-meter dish from bow to stern, with room to spare." You can learn more fast at fast.bao.ac.cn/en/FAST.html.

Hot targets for SETI

SETI projects aren't just focused on random stars; they're also aimed at stars known to have *exoplanets* (the term for planets of stars other than the Sun; see the later section "Discovering Alien Worlds"). After all, we think an alien civilization has to be located on a planet. Specifically, the prime targets are stars with exoplanets that are thought to have solid surfaces (like Earth) with surface temperatures that allow water to exist as a liquid (also like Earth). Many of these exoplanets were found by NASA's Kepler satellite.

Other exoplanets have been revealed with telescopes on the ground, such as the two 24-inch robotic telescopes operating in Chile and Morocco for the TRAPPIST (Transiting Planets and Planetesimals Small Telescope) project. "TRAPPIST" also alludes to the famous beers brewed by Trappist monks in Belgium (Belgium's University of Liège runs the project). For more about TRAPPIST, surf to www.trappist.ulg.ac.be/cms/c_3300885/en/trappist-portail.

SETI wants you!

TIP

Scientists of the SETI@home project seek your help in identifying alien radio signals. Specifically, they want to use your computer from afar when it would otherwise be idle. SETI@home divides a huge data stream from ongoing SETI projects among a great many personal computers of folks like you. If you join SETI@home, your computer will automatically process a small chunk of the SETI data when the machine is idle but turned on. You can go to sleep, but your device may pull an all-nighter.

To join the search, visit the website `setiathome.berkeley.edu` and follow the simple instructions to download free software. From then on, when you're not using the computer, it will sometimes light up with a snazzy SETI screensaver. That's the sign that the computer is crunching away on observational data. Now and then, the computer will connect to a server in Berkeley, California, and send the results of its computing.

In SETI@home's first ten years of operation, more than 5 million people worldwide volunteered the use of idle computer time. As their computers reported back to project headquarters, sometimes they revealed that a suspicious signal was found. SETI scientists examined the reports and, up to now, rejected all of them, meaning that they weren't evidence of E.T. But someday E.T. may call, and your computer may identify the ring. I like the tone of that!

Discovering Alien Worlds

Exoplanets (also called extrasolar planets) are planets that orbit stars other than our Sun. At one time, astronomers didn't know whether a single exoplanet existed, but since the 1990s, they've discovered many of them. The known exoplanets now far outnumber the eight planets of our solar system. As of March 2, 2017, the online Extrasolar Planet Encyclopedia listed 3,586 exoplanets in 2,691 planetary systems. But when you add in other suspected or candidate exoplanets, the list comes to 6,207 planets in 5,071 systems. (A *planetary system* consists of the planets that orbit a particular star. "Solar system" is the name of our planetary system, centered on the Sun.) You can check out the encyclopedia at `exoplanets.eu/catalog`.

In this section, I describe how beliefs about exoplanets have changed over time and the ways in which astronomers now find them. I introduce you to the main types of exoplanets, with typical examples. And I describe astrobiology, the science that investigates the possibility that life exists on some of these alien worlds.

Changing ideas on exoplanets

Scientists and other scholars wondered for centuries whether there were planets around other stars. Because there was no evidence of them until the 1990s, few believed in exoplanets. Of those who did believe, one of the most famous was Giordano Bruno, the Italian Renaissance philosopher. He maintained that the stars in the sky resemble our Sun, long before that was known. He theorized that the stars also have planets that are inhabited, like Earth. These were only some of his unpopular views (he also denied some basic teachings of the Catholic Church and was known as a magician). Bruno was condemned and burned at the stake in 1600.

Long after Bruno's time, and through most of the 20th century, astronomers doubted that many (or even any) exoplanets existed. They thought that the planets of our solar system must have formed when a passing star nearly collided with the Sun. Supposedly, tidal force exerted by the passing star pulled gas out of the Sun, and some of the gas condensed and formed the planets. But a close encounter of this stellar kind is incredibly rare because stars are light-years apart. Few, if any, collisions would mean there are few or no exoplanets.

Astronomers' ideas about the possibility of exoplanets changed in the 1990s, when the Hubble Space Telescope and other instruments revealed that many newborn stars (the Young Stellar Objects that I describe in Chapter 11) are surrounded by disk-shaped clouds of gas and dust. Conditions in these clouds are right for planets to form. In other words, planet birth is a byproduct of common star formation, not rare collisions. But astronomers needed to discover some exoplanets and see whether they actually were as common as this theory implies.

WINSTON CHURCHILL, EXOPLANETARY VISIONARY

One famous nonscientist, British leader Winston Churchill, thought that planets of stars beyond the Sun, as well as life on them, was likely. In 1939, he reasoned on logical grounds that the Sun and the life-bearing Earth should not be unique in the universe and that planets may form in ways that astronomers didn't yet know. He also argued that some of those planets (what we now call exoplanets) would be located at the right distances from their stars to be habitable. His thoughts deserved to be quoted in astronomy textbooks, but they were little noticed or at least soon forgotten.

Over the years, a few astronomers claimed to find exoplanets. But the reports were disproven or could not be confirmed. Success finally came in 1992, when radio astronomers detected two planets of a pulsar. (I describe pulsars in Chapter 11.) Then in 1995, researchers discovered the first exoplanet orbiting a normal star. Many more astronomers joined the hunt, with newer and more powerful instruments. By February 2012, the searchers had found 760 exoplanets and more than 2,000 other possible planets that needed to be confirmed. NASA's Kepler satellite was turning up exoplanets in large numbers, and an expert with the Kepler project estimated that there may be 100 billion planets in our galaxy. As of this writing, the number of confirmed exoplanets is 3,586, and astronomers expect to detect many more.

Finding exoplanets

Exoplanets are much dimmer than the stars they orbit, so a planet is almost always lost in the glare of its star. Therefore, astronomers don't look for planets directly. Instead, they seek telltale effects in the appearance of stars that betray their hidden planets.

The most important clues to the existence of an exoplanet are as follows:

>> **A repeated wobble in the motion of a star:** When astronomers find a star wobbling back and forth repeatedly, with each wobble lasting as long as the previous one, they conclude that it has a faint companion. Gravity makes the unseen companion and the visible star follow orbits around their common center of mass, just like the two members of a binary star (see Chapter 11). But if the companion can't be seen, it may be much smaller and dimmer than a star, and thus a planet. Astronomers have found many such wobbling stars by observing their spectra change with time. The Doppler effect (which I also explain in Chapter 11) reveals the back-and-forth motions.

The *Automated Planet Finder* at Lick Observatory (which I describe in "Space scanning with other SETI projects" earlier in this chapter), uses the wobble method (to scientists, the *radial velocity* technique). It's so sensitive that it can measure a star's wobble speed as slow as the rate of a person walking. At a given distance from a particular star, the smaller (less massive) a planet is, the slower the wobble that its gravity induces in the star. So the smaller the star motions that you can measure, the smaller the planets that you can detect, other things being equal.

>> **Periodic dimming of a star:** When observers make precise measurements of the brightness of a star and detect a very small dimming, it may mean that the star has a planet that is passing across the face of the star, just like a transit of Mercury across the sun (see Chapter 6). The planet blocks a little of the star's

light while it passes in front. The fraction of the star's light that's missing during a transit tells astronomers the size of the planet, compared to the star. (The bigger the planet, the larger the drop in brightness.) The interval between two transits is the time it takes the planet to make one orbit around the star: the planet's "year." Kepler and a French satellite called CoRoT look or looked for exoplanets by this transit method, and I list a few of these finds and their years in Table 14-1. (CoRoT operated from 2006 to 2012; Kepler was launched in 2009 and is still working, but it's not quite as good as new.)

The TRAPPIST telescopes in Chile and Morocco (see "Hot targets for SETI" earlier) and other ground-based telescopes are finding exoplanets by the transit method, and so will the Transiting Exoplanet Survey Satellite (TESS), which NASA plans to launch in 2018. CHEOPS (Characterising Exoplanet Satellite), a satellite developed by the European Space Agency and the Swiss Space Office, will observe some exoplanets that were found by the wobble method, to detect periodic dimming and thereby measure the planets' diameters. If all goes well, CHEOPS will launch in 2018.

» **A brief increase in star brightness, followed (or preceded) closely by another fleeting increase:** Some astronomers use telescopes that monitor thousands of stars at once. They're looking for unusual events, when one of the stars brightens noticeably and then fades to its original magnitude over a period of a few weeks. During those weeks, there may be a second brightening lasting just a few hours or days. The brightenings are the result of gravitational microlensing (which I explain in Chapter 11). In this process, the gravity of a dim foreground star magnifies the light of the more distant star for a few weeks, making it appear brighter. When a second, briefer brightening takes place, that additional effect is caused by the gravity of a planet near the foreground star.

The Optical Gravitational Lensing Experiment (OGLE) has discovered six exoplanets by the microlensing technique. Operated by the University of Warsaw in Poland, it uses a 51-inch telescope at Las Campanas Observatory in Chile. When you monitor many stars simultaneously, you also detect exoplanet transits, and OGLE has found 33 of them. Yet its intended purpose was to study dark matter (which I discuss in Chapter 15).

Planet mass is listed in units of Earth's mass, M_E, or Jupiter's mass, M_J. Size is given in terms of the diameters of Earth or Jupiter, D_E and D_J. A "?" means that the quantity is unknown. The length of a planet's "year," meaning the time it takes to go once around its star, is stated in units of an Earth day, an Earth hour, or an Earth year. You can read about some of the designations from the Description column (such as Exo–Earth) in the following section.

TABLE 14-1 **Notable Exoplanets**

Planet	Mass	Size	"Year"	Distance	Description
PSR 1719-14 b	1 M$_J$	0.4 D$_J$	2.2 hours	3,900 light-years	Pulsar planet, diamond planet
Kepler-138 b	0.07 M$_E$	0.6 D$_E$	10 days	220 light-years	Exo-Earth
Kepler-20b	9.7 M$_E$	1.9 D$_E$	3.7 days	950 light-years	Very massive Super-Earth
GJ 1214b	6.6 M$_E$	2.7 D$_E$	1.6 days	47 light-years	Mini-Neptune
55 Cancri e	8.3 M$_E$	2.0 D$_E$	18 hours	40 light-years	Rocky Super-Earth, probably highly volcanic
J 1407b	20 M$_J$	?	10 years	430 light-years	Massive Jupiter with huge rings
CoRoT-9b	0.8M$_J$	0.9 D$_J$	95 days	1,500 light-years	Garden-variety Jupiter
WISE 0855—0714	6 M$_J$	?	Not applicable	7 light-years	Interstellar (rogue) planet
Kepler-16b	0.3 M$_J$	0.8 D$_J$	229 days	200 light-years	Tatooine planet, orbiting two cool dwarf stars
WASP-17b	0.5 M$_J$	2.0 D$_J$	3.7 days	1,000 light-years	Wrongway planet, hot Jupiter
Gliese 436b	22 M$_E$	4 D$_E$	2.6 days	33 light-years	Hot Neptune with giant gas tail
TRAPPIST-1 f	0.7 M$_E$	1.0 D$_E$	9.2 days	39 light-years	Goldilocks Exo-Earth

The vast majority of exoplanets are found by the wobble, transit, and microlensing observations. But two other methods have found small numbers of unusual exoplanets:

>> **Direct imaging:** Sometimes, astronomers find exoplanets that are *not* lost in the glare of their stars. The planets show up as tiny dots alongside the stars in telescopic photos. Photos taken over a period of time show whether the "dot" moves through space with the star, confirming that it's a planetary companion, not a faint background object. The Gemini Planet Imager (GPI), a powerful camera and spectrograph used with the 320-inch Gemini SouthTelescope atop Cerro Pachon in Chile, operates in infrared light. Young Jupiter-class exoplanets still retain heat from their formation and glow in the infrared. In 2015, the GPI discovered such a planet, 51 Eridani b.

Further advances in direct imaging are underway. One big step is the modification of an existing instrument used for studies in infrared light at one of the four 8.2-meter (323-inch) reflectors of the Very Large Telescope (VLT), also in Chile. Scientists are upgrading the instrument to hunt for planets of Alpha Centauri A (Rigil Kentaurus) and Alpha Centauri B, two of the three nearest stars beyond the Sun. It will be provided with adaptive optics (which improves the image sharpness by compensating for disturbances in the air above the telescope that cause blurring) and a coronagraph. Coronagraphs block the light from a star to make it easier to see faint objects near it. A similar instrument will be mounted on the future largest telescope in the world, the 39-meter (128-foot) European Extremely Large Telescope (E-ELT). (You can see live images as a webcam pans around the VLT observatory by heading to www. eso.org/public/usa/teles-instr/paranal/.) Both VLT and E-ELT are projects of the European Southern Observatory.

>> **Pulsar timing:** Radio astronomers measure the arrival times of pulses of radio waves from pulsars, a kind of dead star that I describe in Chapter 11. The pulses come at precisely spaced intervals. But in a few cases, pulses arrive ahead of schedule for a while and then behind schedule for an equal period of time, and that pattern keeps repeating. This pattern reveals that the pulsar is going around a small orbit because of the gravity of a planet. When the pulsar is on the side of the orbit that's closer to Earth, the pulses arrive early because they have a slightly smaller distance to travel. And when the pulsar is on the far side, pulses come late because they have a longer way to go. Most exoplanets were born along with their stars, but pulsar planets probably formed after their stars died in supernova explosions. Planets around pulsars are extremely rare, but they were the first planets discovered outside our solar system.

Meeting the (exo)planets

Astronomers probably haven't even found all kinds of exoplanets because some are too small or too rare to show up in current observations. Current knowledge of exoplanets is thus likely incomplete. Yet astronomers have identified many interesting kinds already, most of them unlike any planet in our solar system.

Take a look at the main types of known exoplanets:

>> **Carbon planet:** A rocky world with a lot more carbon and much less silicate rock and water (if any) than Earth. Its surface layer may be mostly graphite (like the "lead" in a pencil); underground, the carbon forms a layer of diamond (making it also a *diamond planet*).

>> **Exo-Earth (or Earth, for short):** A rocky planet with about the same size and mass as our Earth.

>> **Super-Earth:** An exoplanet that's bigger and more massive than an exo-Earth, but smaller than Neptune, more or less. Super-Earths range in mass from twice to ten times the mass of Earth. A Super-Earth may be a rocky planet; a gassy, icy planet like a much smaller Uranus or Neptune (see Chapter 9); a carbon planet; or even a water world (described later in this list).

>> **Goldilocks planet:** An Exo-Earth, or a rocky Super-Earth, with surface conditions suitable for liquid water to exist. It should be located in the *habitable zone,* which is the range of distances from its star within which surface water neither freezes permanently nor boils away. Some rocky exoplanets with thick atmospheres may be in the habitable zone but aren't Goldilocks planets because their atmospheres trap so much heat that the surface temperature is far above the boiling point. (That description sounds just like Venus, which I describe in Chapter 6.)

>> **Hot Jupiter (also called a *roaster*):** A gas giant planet like Jupiter (see Chapter 8) that's very close to its sun. Many hot Jupiters are closer to their suns than Mercury is to ours. See Figure 14-3 for an artist's rendering of a hot Jupiter exoplanet.

>> **Jupiter:** A gas giant exoplanet that's located far enough from its Sun that it's cold, like our Jupiter. I call this kind of exoplanet a "garden-variety exoplanet," although, so far, astronomers have found far more hot Jupiters than cold ones. A Jupiter that's far from its sun may be hotter than our Jupiter if it's younger and hasn't cooled off as much yet.

>> **Rogue or interstellar planet:** An exoplanet that isn't orbiting a star. It may have been ejected from an orbit around a star, or it may have formed on its own from a cloud of interstellar gas and dust, like stars form.

>> **Tatooine planet:** An exoplanet that orbits a binary star so it has two suns. The name comes from the fictional planet Tatooine, with two suns and a desert landscape, that's Luke Skywalker's home in the *Star Wars* films.

>> **Tidal Venus:** An Exo-Earth that, although it may be located in a habitable zone, is (like Venus itself) much too hot to have liquid water. Our Venus is hot from trapping solar energy in the thick atmosphere. But a Tidal Venus is heated by strong tidal force exerted by its sun, which causes friction in its rocky interior.

>> **Water world:** A Super-Earth that's substantially made of water. Compare that with our Earth, which is made of rock and iron, with oceans on the surface. A water world can be half or more made of water, with no place to dock your boat.

>> **Wrongway (retrograde) planet:** An exoplanet that orbits its sun in the opposite direction to the way the star turns. In the solar system, all eight planets (and Pluto, too) follow *prograde* orbits. In other words, they orbit in the same way our Sun turns: counterclockwise, as seen from an imaginary point far above the north pole of the Sun. A retrograde planet is like the famous aviator Douglas "Wrong Way" Corrigan, who flew from New York to Ireland in 1938 when he was supposed to head for Long Beach, California.

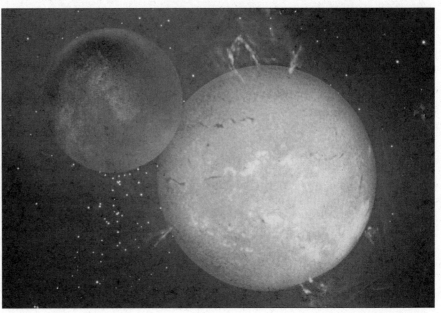

Courtesy of Seth Shostak

FIGURE 14-3:
An artist's concept of a hot Jupiter orbiting the star 51 Pegasi.

I list typical examples for several types of exoplanet in Table 14-1. Many of the known exoplanets are part of larger systems, like TRAPPIST-1 with its six known fellow orbiters.

Catching Proxima fever: Focusing on red dwarfs

The August 2016 announcement of an exoplanet orbiting the red dwarf star Proxima Centauri thrilled astronomers and space enthusiasts of all kinds. Here's why:

» Proxima Centauri (or Proxima Cen) is the nearest star to our solar system, only 4.24 light-years from Earth.

» The planet, called Proxima b, is quite possibly Earth-like, with a mass that may be as small as just 30 percent more than Earth's mass.

» Proxima b is located in the star's habitable zone, the range of distance from Proxima Cen where water can possibly exist as a liquid on the surface of a rocky planet.

In other words, Proxima b may be an Earth–like planet with conditions suitable for life. But that conclusion isn't a done deal for the following reasons:

>> Although Proxima b is in the habitable zone, its distance from the star is only 5 percent of Earth's 93-million-mile distance from the Sun. Its orbital period or "year" is only 11.2 Earth days, and it very likely keeps one and the same side always facing its star while the opposite side of the planet never faces the star. (That's like our Moon, which has the "near side" that always faces Earth and the far side that always faces away from us, as I tell in Chapter 5.)

>> Proxima Cen is a flare star (see Chapter 11) and has starspots and other stellar activity in a 7-year magnetic cycle, reminiscent of the Sun's 22-year magnetic cycle of solar activity, sunspots, and more. (See Chapter 10.)

>> Calculations suggest that Proxima Cen has a stellar wind, similar to the Sun's solar wind but more powerful. (See Chapter 10.)

>> If Proxima b's mass turns out to be somewhat larger than 1.3 Earth masses, it may not be a rocky planet but rather a gaseous one, like a smaller version of Neptune (see Chapter 9 for all about Neptune).

What those four statements add up to is the possibility that Proxima b is any one of the following:

>> A balmy hospitable planet with temperatures in the 80s (Fahrenheit; 26° to 32° Celsius) and plenty of water for life (should life arise), with a nice atmosphere that shields the planet's surface from much of the dangerous ultraviolet light and X-rays that come from flares on Proxima Cen, and a strong magnetic field that produces a magnetosphere. The atmosphere would also distribute heat, in winds, from the star-facing side to the opposite hemisphere, which would otherwise be very cold. The magnetosphere would shield the planet from most of the powerful stellar wind that otherwise would sweep away the atmosphere and bombard the surface. (I tell how Earth's magnetosphere protects us in Chapter 5.)

>> A cold, rocky world with no liquid water on the surface and not much of an atmosphere like Mars (see Chapter 6).

>> A rocky, horribly hot world with a thick atmosphere and strong greenhouse effect, like Venus (see Chapter 6).

>> A gaseous world like a mini Neptune. (I describe Neptune in Chapter 9; there is no miniature Neptune in the solar system, although exoplanets of that type have been found orbiting some stars in the Milky Way.)

Only further research will tell us for sure which of these possibilities is correct, but that work won't be easy. If we could observe transits of Proxima b in front of Proxima Cen by using the periodic dimming method I explain earlier in the chapter, that would be wonderful. Then we would know both the size and mass of the exoplanet, calculate its density, and see what kind of planet that implies. But only a tiny fraction of all exoplanets in the galaxy have orbits that are oriented so that the planets pass directly in front of their stars as seen from Earth. If they don't do so, we can't see them transit. Estimates indicate that we have a less than a 2 percent chance to see Proxima b in transit. So far, no astronomer has reported observing a transit of this exoplanet.

One way to learn more about Proxima b or other possible planets of Proxima Cen or its neighbor stars Alpha Centauri A and B is to send a probe to the Alpha Cen system (Proxima is probably part of the Alpha Centauri system — at least, it's close to it — and is also known as Alpha Cen C.) As I mention in the first section of this chapter, a rocket of the kinds we have now would take about 1,000 centuries to reach Alpha Cen, and none of us can wait that long. However, a much faster means of interstellar travel is in development. See the "Inventing new methods of interstellar travel" sidebar for details.

Finding Earth-class planets orbiting TRAPPIST-1

Another planet-bearing red dwarf star has got astronomers excited. The star, TRAPPIST-1, isn't as close as Proxima (39 light-years versus 4.24), but it has at least seven Earth-class planets, several of which could have liquid water! I list one of them at the end of Table 14-1.

When Frank Drake kicked off the modern SETI programs, he and other experts concentrated on stars like the Sun because the only known planets were the planets of our Sun, and the only known life (intelligent or otherwise) was on one of them. Now it seems clear from Proxima, the seven TRAPPIST-1 planets, and other examples that red dwarfs are likely to have Earth-class planets that may be habitable. Astronomers need to spend more time investigating them.

If there are advanced civilizations on two or more of the TRAPPIST-1 planets (I'm not saying that's likely), they could be visiting, trading with, or even attacking each other right now. Hopefully they won't join forces to attack Earth. But I won't believe any of that until we make contact with them or vice versa. (SETI scientists are listening in the TRAPPIST-1 direction.)

INVENTING NEW METHODS OF INTERSTELLAR TRAVEL

A new initiative, *Breakthrough Starshot,* has started work on a revolutionary method of interstellar travel that that may be capable of sending probes to fly past Proxima b on a journey of just 20 years. (Its initial funding comes from billionaire philanthrophist Yuri Milner, who's also paying for the *Breakthrough Listen* project I describe in the earlier section "Space Scanning with other SETI projects.") Sorry, you can't hitch a ride. Each tiny probe would be a high-tech chip less than an inch wide — perhaps like one in your smartphone but much more advanced, able to take pictures and possibly beam its scientific measurements back to Earth. (Don't even think about the roaming charges.)

A rocket would be used just to lift hundreds of Starshot probes above Earth's atmosphere and release them. Each chip would be attached to a "light sail," a perhaps 13-foot-wide piece of an incredibly lightweight and almost perfectly reflecting material that still needs to be invented. Then an extremely powerful set of lasers on Earth would shine on the light sails for a brief while, propelling them as the laser light bounces off the sails and accelerates them to 20 percent of the speed of light (meaning to 37,200 miles per second). Then the lasers would turn off and the probes would keep going at that fantastic speed.

Many striking advances in technology and materials must be accomplished to make Starshot work. But a significant number of the best science and engineering minds have signed on to help. It will take at least 20 years to get ready to launch, 20 more years to reach Proxima b, and a bit over four more years for the pictures and data to travel back to Earth. I hope that many of the younger readers of *Astronomy For Dummies* will still be around circa 2061 to greet the pictures from Proxima. If Starshot doesn't succeed by then (some think it will take 40 or even 50 years before all technical challenges are met and the probes can be launched), they can at least wait another year and see Halley's Comet when it returns in 2062.

Checking out planets for fun and science

TIP

You can check out the Kepler satellite and its planetary discoveries at the NASA website `kepler.nasa.gov`. Consult the Open Exoplanet Catalogue every so often for the latest info on the number of exoplanets and their individual properties (`www.openexoplanetcatalogue.com`). Another excellent source is the NASA Exoplanet Archive, `exoplanetarchive.ipac.caltech.edu`.

You'll find well-written reports on recent exoplanet discoveries at Sky & Telescope's exoplanet news page, `www.skyandtelescope.com/astronomy-news/exoplanets/`. There aren't quite as many good exoplanet websites as there are confirmed exoplanets, but there are lots of both.

Many folks like you have contributed to research on exoplanets by joining the Citizen Science Planet Hunters project maintained by Yale University and Zooniverse at the website www.planethunters.org. They took an active role in investigating the data from Kepler. As this book went to press, the Planet Hunters project was temporarily closed until enough new data are available for Citizen Scientists to study. So check their website from time to time to see when they need you.

Astrobiology: How's Life on Other Worlds?

Astrobiology is the science that studies the possibility of life in space. If the galaxy has billions of exoplanets, at least some should have the right conditions for life to arise and flourish. Thus far, scientists have found only a tiny fraction of the billions of planets, and even the closest is light-years away. We can't get a closeup picture of an exoplanet with a big telescope and look for herds of dinosaurs or the Great Wall of Centaurus.

The two most fruitful avenues of examining the possibility of extraterrestrial life right now are as follows:

>> Studying *extremophiles,* meaning life forms that exist in extreme environmental conditions on Earth — conditions that are deadly to most ordinary kinds of life but that may prevail on another world

>> Searching nearby bodies in the solar system with space probes and telescopes for signs of life, current or past

I describe how researchers use both methods in the following sections.

Extremophiles: Living the hard way

Most extremophiles are microorganisms, like bacteria, although many species of plants survive at temperatures below 0°F in Antarctica. If you enjoy extreme cold, you're a *cryophile.* Some bacteria flourish in water so hot that it would scald to death everything else. These *hyperthermophiles* live at up to 200°F in hot springs (like the famous ones in Yellowstone National Park) and at even higher temperatures in deep ocean vents. The water temperatures can climb above the normal boiling point, 212°F, but water that hot doesn't boil when under high pressure, deep below the ocean surface.

Some scientists think that life on Earth may have begun near deep sea hydrothermal vents like those found in the present-day oceans. The world's oldest likely fossils, reported in 2017, appear to be tiny structures resembling those often formed by bacteria at deep vents today.

The hyperthermophiles we know of couldn't last on Venus, where the surface temperature is 870°F (see Chapter 6). But there's probably a moderately hot exoplanet somewhere that's just right for organisms like them.

If you're a gardener, you may check the pH of your soil to make sure it's not too alkaline or acidic for the plants you want to grow. Too much alkali (or acid) can be fatal for most living things. But certain extremeophiles love ultra-alkaline conditions (alkalophiles) or swim through acid waters with impunity (acidophiles). Drop a fish in the Dead Sea, and it dies (hence, the name). But for some bacteria (halophiles), that salty sea is sweet indeed.

Some bacteria live in tiny pores in solid rock more than 3 miles (5 kilometers) underground. Those extremeophiles don't get energy from the Sun; they run on chemical energy from their immediate environment. One species that lives deep in a mine even gets its energy from the radioactive decay of uranium in the rock. And barophiles thrive in the ocean depths, where the pressure of the overlying water is a thousand times the atmospheric pressure at sea level. Scientists have even discovered bacteria that live in the clouds above your head and strange cave-dwelling microbes unlike any known bacteria.

What extremeophiles tell us is that life is opportunistic. It will find ways to survive — and perhaps to arise in the first place — in conditions under which we can't exist. What can happen on Earth can happen on exoplanets. And that's just assuming that life on other planets is like life as we know it. If other life is completely unlike what we know — for example, life that's not based on carbon (like Earth life), but on another element — all bets are off, and many otherworldly environments may be inhabited.

Seeking life in the solar system

Obtaining reliable evidence of life off Earth is going to be difficult. But we'll never know if we don't try. Astronomers have identified the following locations as the best places to look for life in the solar system (outside Earth):

» Mars

» Jupiter's moon Europa

» Saturn's moons Titan and Enceladus

The ongoing search for Martians

As I describe in Chapter 6, scientists examined and disputed the claim that microscopic fossils are present in a rock from Mars. Astronomers studying Mars with telescopes on Earth get conflicting results about whether methane gas appears from time to time on Mars. One source of methane could be bacteria (some species on Earth make methane, and others eat it). But even if methane does exist on Mars, it may come from a geological process, just as some gases on Earth come out of volcanoes. NASA space probes mapped Mars landforms that look like the remains of ancient sea beds and flood channels. Robotic Mars rovers found rocks and minerals that formed when pools of water dried up. Where there was water on Mars, there may have been life. Much of Mars has permafrost. Below the permafrost, where it's warmer underground, there may be liquid water. Perhaps microbes are present on Mars now, living below the reach of the rovers' soil scoops. Inquiring minds want the scoop.

The most recent robotic rover to traipse across the red planet, NASA's *Curiosity*, landed in 2012 and is still exploring. *Curiosity* is heavily instrumented to analyze chemical and geologic properties of the areas that it visits in search of clues to whether an ancient surface where water once was abundant would have been hospitable to microbial life. At a place named Yellowknife Bay, where flowing water once deposited rocky debris and sediments, the rover found

>> Clay soils, considered favorable for life to arise according to some theories

>> A pH of the ancient water suitable to support life

>> At least some chemicals needed for life, including soil nitrates

Curiosity finds traces of methane on Mars from time to time, but we can't tell where the methane comes from. In any case, I conclude that Mars could have supported life on the surface at one time, perhaps 4 billion years ago. But we have no evidence that it did.

Europa, we hardly know thee

Europa is a rocky Jupiter moon with a surface layer of ice (see Chapter 8). Liquid water (probably salty) covers the whole moon under the ice. Scientists believe the ocean could host microscopic life. However, the ice may be 10 miles (16 kilometers) thick. We may not have the engineering capability to drill down to the ocean on Europa anytime soon, even if NASA or another space agency had the money to try it. Some Europa experts suspect that there's occasional turnover in the ice and ocean. If so, some water gets to the surface, where it freezes and remains. If that's true, the upheavals may bring traces of ocean life to the surface, where NASA's future lander probe (still in the planning phase) may someday find it.

MOONING OVER EXTRATERRESTRIAL LIFE

The fact that moons like Europa, Titan, and Enceladus may be capable of supporting life has an important implication for astrobiology: When astronomers think about life on exoplanets, they need to consider the moons as well. Astronomers are starting to search for exomoons because where there are exoplanets, there should be some moons, too. Some exomoons have been reported but aren't yet confirmed. If a planet is in a habitable zone but is itself unsuitable for life (because its atmosphere traps too much heat, for example), it may have a moon that's just right for life in the zone.

Titan: Perhaps a primeval Earth

Titan, the largest moon of Saturn, is bigger than Earth's Moon and has a thick atmosphere, making it more like a planet than any other moon (see Chapter 8). It also has large lakes of liquid hydrocarbons. Astronomers think Titan may resemble the very early Earth, before oxygen became an important part of the atmosphere. So if life arose on Earth, it could happen on Titan. Titan is colder than Earth ever was, but it does have an underground water ocean. Some experts theorize that a new kind of life may form in the hydrocarbon lakes. So scientists are studying the closest thing to a Titan lake on Earth, the Pitch Lake of liquid asphalt in Trinidad. (For more on Titan, turn to Chapter 8.)

The Enceladus plume

Enceladus is an icy moon of Saturn that has a body of water beneath the ice, at least near the south pole. Quantities of water spout out from the polar region into space, freezing immediately into tiny ice particles. Unlike on Europa, where the ocean is deep beneath thick ice, Enceladus's water is close to the surface. Because of the water's proximity to the surface, scientists may find it easier to sample and look for evidence of life.

Dr. Seth Shostak, senior astronomer at the SETI Institute in Mountain View, California, contributed this chapter to earlier editions of Astronomy For Dummies. *The author, Stephen P. Maran, revised and rewrote the chapter for this edition. All opinions expressed in this chapter are those of the author.*

Chapter **15**

Delving into Dark Matter and Antimatter

S tars and galaxies set the night sky aglow, but these glittering jewels account for only a tiny portion of the matter in the cosmos. There's more to the universe than meets the eye — much more.

This chapter introduces you to the concept of dark matter, tells you why astronomers are convinced that the stuff exists, and describes experiments that may shed light on the nature of this mysterious, invisible material. I also discuss another exotic type of matter in the universe: antimatter. Yes, antimatter exists outside of science fiction, and the real-world version is every bit as fascinating as the sci-fi books, television shows, and movies suggest.

Dark Matter: Understanding the Universal Glue

As far back as the 1930s, an astronomer found hints that the vast majority of the mass in the universe doesn't emit, reflect, or absorb light.

The invisible material, known as *dark matter*, serves as the gravitational glue that keeps a rapidly rotating galaxy from flying apart and enables fast-moving galaxies in a cluster to stick together. Dark matter also seems to have played a crucial role in the development of the universe as we know it today — a spidery network of immensely long superclusters of galaxies separated by giant voids (see Chapter 12).

Astronomers have determined that almost 85 percent of all the matter in the universe is dark matter. What a humbling thought. The universe you see when you peer through a telescope or look up at the night sky teeming with stars and galaxies is just a tiny fraction of what's out there. To borrow a nautical analogy, if galaxies are like sea foam, dark matter is the vast, unseen ocean on which it floats.

Gathering the evidence for dark matter

The first hint that the universe contains dark matter appeared in 1933. While examining the motions of galaxies within a large cluster of galaxies in the constellation Coma Berenices, astronomer Fritz Zwicky of the California Institute of Technology found that some galaxies move at an unusually high speed. In fact, the galaxies of the Coma Cluster move so rapidly that all the visible stars and gas in the cluster can't possibly keep the galaxies gravitationally bound to one another, according to our known laws of physics. Yet somehow the cluster remains intact. (The Coma Cluster is about 320 million light-years from Earth. To learn more about clusters of galaxies, check out Chapter 13.)

Zwicky concluded that some sort of *dunkle Materie* ("dark matter" in German, the language in which his work was originally published) must exist in the Coma Cluster to provide the missing gravitational attraction.

Scientists often fail to appreciate an astonishing discovery when just one person or team claims it. They want to see more evidence, from independent experts, before they're convinced of the new findings. So it's not surprising that the concept of dark matter wasn't accepted until decades after Zwicky's work. Many astronomers ignored his report or figured that after they studied the motions of galaxies in greater detail, the rationale for the existence of the supposedly invisible material would disappear.

In the 1970s, astronomers began to find additional and compelling evidence for dark matter. Not only do clusters of galaxies seem to contain the dark stuff, but dark matter is also in individual galaxies. The following sections describe the main arguments in favor of dark matter.

Dark matter makes stars orbit oddly

Vera Rubin and Kent Ford of the Carnegie Institution of Washington, D.C., were studying the motions of stars in hundreds of spiral galaxies with a powerful new spectrograph when they obtained a result that seemed to fly in the face of conventional physics. (A *spectrograph* is an instrument that separates the light from a star or other source into its constituent colors or wavelengths.) A typical spiral galaxy resembles a flattened fried egg, with most of its visible mass (stars and bright nebulae) seemingly concentrated in the yolk — astronomers call this the *bulge* (as I explain in Chapter 12). Images reveal that the visible mass of a spiral diminishes rapidly with increasing distance from the bulge.

Astronomers presumed that the stars in a spiral galaxy orbit around the galaxy center in the same way that the planets in our solar system orbit the Sun. Obeying Newton's law of gravity, the outer planets, such as Uranus and Neptune, orbit the Sun less rapidly than the inner planets, such as Mercury and Venus. Therefore, stars on the outskirts of a spiral galaxy should orbit at a lower speed than stars nearer the center. But that's not what Rubin and Ford found.

In galaxy after galaxy, their observations revealed that the outlying stars orbit rapidly, just like the inner stars. With so little visible material in the outer regions, how do the outlying stars manage to zip around so fast and still stay bound to the galaxy? They should escape from their galaxy, given their speed. Rubin had found serious hints of this unexpected behavior in earlier work, but many astronomers needed to be convinced of it. (For more about escape velocity, see Chapter 13.)

ASTRONOMERS WHO SHED LIGHT ON DARK MATTER

Astronomers doubted the evidence for dark matter that Fritz Zwicky and Vera Rubin found decades apart, but today few experts doubt it. Both of these pioneering astronomers have passed away, Zwicky in 1974 and Rubin in 2016. Zwicky also contributed to the study of supernovas (indeed, he coined the word) and cosmic rays.

Rubin made many contributions to the understanding of galaxies and, having experienced discrimination against women in observatories early in her career, became a mentor and role model for younger women astronomers and a leader in the campaign for women's equality in the profession. At the time of her death, a petition had been circulating urging that she receive the Nobel Prize in Physics for her contributions to the discovery of dark matter, but unfortunately that won't happen because the Nobel Prize is not given posthumously.

Kent Ford developed the electronic sensor or image tube that was the key component of the Image Tube Spectrograph (ITS) used for his and Rubin's discovery. The ITS was later named one of the 101 Objects That Made America in the collections of the Smithsonian Institution. You can see it on exhibit at the National Air and Space Museum when you visit Washington, D.C.

After the findings of Rubin and Ford became well known, astronomers concluded that *visible matter* — the stars and luminous gas that show up on telescopic photographs — makes up only a small portion of the total mass of a spiral galaxy.

Although the visible mass is indeed concentrated toward the center, a vast quantity of other material must extend far beyond. Each spiral galaxy is surrounded by an immense *halo* of dark matter. And to exert enough of a gravitational tug on the stars in the visible outskirts of the galaxies to make them orbit as rapidly as observed, the dark matter must exceed the visible matter by a factor of 10 in mass. Other types of galaxies, including elliptical and irregular galaxies, also have dark-matter halos. Dwarf galaxies (which I describe in Chapter 12) have even higher proportions of dark matter to visible matter than large galaxies.

The Milky Way galaxy is about 100,000 light-years in diameter (that's the spiral-shaped disk of stars). But the halo of dark matter around the Milky Way, shaped like a sphere, is at least 600,000 light-years across. That means that some much smaller nearby galaxies, including the Large Magellanic Cloud (distance: about 163,000 light-years) and the Small Magellanic Cloud (200,000 light-years) are inside the dark halo. (I describe those galaxies in Chapter 12.)

Starting in the 1990s, astronomers who observed clusters of galaxies with X-ray telescopes on satellites such as ROSAT and the Chandra X-ray Observatory mapped huge regions in and around the clusters that glow in X-rays. The glows come from an *intracluster medium* of thin, very hot gas. The hot gas fills such a huge region that, thin as it is, its total mass exceeds the sum of the masses of all the galaxies in a cluster.

The thin, hot gas would expand away, but it's kept in place at the cluster of galaxies by the gravitational attraction of a mass that is much larger than its own mass plus the mass of the cluster galaxies. The source of the powerful gravity is dark matter in the cluster. This fact is further evidence that Fritz Zwicky was right in 1933 when he claimed that unseen matter (which we now call dark matter) exists in huge amounts in a cluster.

Dark matter causes gravitational lensing

Dark matter is also revealed by *gravitational lensing*, the bending of light by large masses in space, as predicted by Albert Einstein. Astronomers observe this effect when

>> They find two or more quasars close together that turn out to be one and the same quasar, with its light bent and focused so we see double (or more) from Earth.

>> They find a glowing circle (called an *Einstein ring*) or portions of circles (arcs) that are caused by the gravitational lensing of distant quasars or galaxies by a large mass or masses in the foreground. The rings and arcs are distorted images of

the background objects. The mass that does the bending is usually the dark matter in a cluster of galaxies (the ordinary matter in the cluster contributes just a little to the bending because it has much less mass than the dark stuff).

» They find oddities in the sizes, shapes, and orientations of many galaxies across a region of the sky. By oddity in orientation, I mean a slight tendency of galaxies to line up as though tilted in the same direction. Astronomers don't know the true exact shape, size, and orientation of any one galaxy, but they detect these effects statistically. The effects are called *weak lensing*. Weak lensing isn't caused by a single large mass like the dark matter in a cluster of galaxies but by the cumulative, often individually small effects of any and all matter in space between us and the galaxies under observation.

In other words, when you look far out in the universe, everything you see is distorted at least slightly (weak lensing). Sometimes it's greatly distorted by strong lensing, such as when a quasar that should look like a bright point instead appears as an Einstein ring. When astronomers look at the distant universe, nothing is exactly what it seems; it's all warped, a little bit or a great deal, by gravitational lensing from dark matter.

Cold dark matter makes the universe take its lumps

Cosmologists (scientists who study the large-scale structure of the universe and its formation) also point to dark matter to explain a fundamental puzzle about the universe: How did it evolve from a nearly uniform soup of elementary particles in the aftermath of the Big Bang (see Chapter 16) to reach its present lumpy structure of galaxy clusters and superclusters?

Even though 13.7 billion years have passed since the birth of the universe, scientists don't believe that enough time has elapsed for visible matter to coalesce on its own into the huge cosmic structures we see today.

To solve this cosmological conundrum, some experts state that the universe contains a special type of dark matter, called *cold dark matter,* that moves more slowly but gathers into clumps more quickly than ordinary, visible matter. Responding to the tug of this exotic material, the ordinary matter formed stars and galaxies within the densest concentrations of the dark matter. This theory explains why every visible galaxy seems to be embedded in its own dark-matter halo.

Is the cold dark matter theory correct? It appears to be in broad agreement with the facts about the universe, as far as scientists know them. But the agreement isn't perfect. For example, the theory predicts that hundreds of tiny satellite galaxies will surround a big galaxy like the Milky Way. But we don't see all that many satellite galaxies. The predictions of the theory may need work, or we may need a better theory of dark matter itself. Or maybe there are small, faint galaxies all around us that we haven't found yet.

It's even possible that astronomers haven't found enough satellite galaxies because many of them were swallowed by the large galaxies that they orbit, like the Sagittarius Dwarf Galaxy and the Canis Major Dwarf Galaxy, which are being absorbed now by the Milky Way (as I tell in Chapter 12).

Dark matter is critical to the universal density

Astronomers believe in dark matter for yet another cosmic reason: On large scales, the universe looks the same in all directions and has an overall smoothness. This consistency in appearance and smoothness indicates that the universe has just the right density of matter, called the *critical density* (which I explain in Chapter 16). The total amount of visible matter that we observe in the universe isn't nearly enough to achieve critical density. Dark matter takes up the slack.

Debating the makeup of dark matter

Okay, astronomers have plenty of good reasons to believe in dark matter, as I describe in the preceding sections. But what the heck is the stuff anyway? Broadly speaking, astronomers divide the possible kinds of dark matter into two classes: baryonic dark matter and oddball dark matter.

Baryonic dark matter: Lumps in space

Some dark matter may consist of the same stuff that the Sun, planets, and people are made of. This kind of dark matter would be part of the family of *baryons,* a class of elementary particles that includes the protons and neutrons found in the nuclei of atoms. But it's not the cold dark matter that I mention earlier in the chapter.

Baryonic dark matter includes all hard-to-see material that's made of known kinds of matter, including asteroids, brown dwarfs, and white dwarfs (I describe the dwarfs in Chapter 11). Yes, scientists can detect asteroids in our solar system and white dwarfs and brown dwarfs nearby in the Milky Way. But far out in the galactic halo, these objects may be undetectable with present equipment. According to one theory, such hypothetical objects, called *MACHOs (massive compact halo objects),* could account for the dark-matter halos surrounding individual galaxies if the MACHOS are there in sufficient numbers. (I cover the search for MACHOs in the halo of our galaxy later in this chapter.) But we don't see nearly enough of them to account for the dark halo of the Milky Way, so they probably don't make up the dark matter in other galaxies either. I think this theory is wrong, and the scientists who proposed it will have to take their lumps.

Oddball dark matter: Stranger still

Alternatively, dark matter may consist of one or more types of exotic subatomic particles that bear little or no resemblance to baryons. These particles include *neutrinos,* which do exist (see Chapter 10 for more about them), and others with names such as *axions, squarks, photinos,* and *neutralinos* that physicists have dreamt up without proof of their existence. Experiments are underway, but no one has captured an axion or any other hypothetical dark matter particle thus far — at least, they haven't done it to the satisfaction of other scientists, so we are still in the dark about dark matter.

During the Big Bang at the birth of the universe (see Chapter 16), a zoo of weird dark matter particles may have formed, and some types of them may still exist. The theorized particles include the axion, a kind of miniature black hole 100 billion times lighter than an electron. Even though axions (if real) are featherweights, enough of them can contribute significantly to the cosmic mass. Recent experiments suggest that the neutrino (a particle that scientists once thought may have zero mass) does have a very small but real mass. So neutrinos may account for a small portion of the dark matter.

Other candidates for oddball dark matter are heavier — about ten times the mass of the proton — but still insubstantial in terms of making up the dark matter in the universe, unless they occur in large numbers. These include the yet-to-be-detected partners of certain known subatomic particles such as quarks and photons. These hypothetical dark matter counterparts are squarks and photinos. Many theories exist for these dark matter particles, along with fanciful names for them. Scientists describe them collectively as *weakly interacting massive particles,* or *WIMPs* (covered later in this chapter).

Taking a Shot in the Dark: Searching for Dark Matter

Around the world, physicists are planning or operating sensitive detectors to find the elusive, telltale signals of dark matter. Some of these detectors are designed to analyze the subatomic debris created by giant atom-smashing devices, which briefly reproduce the extreme heat, energy, and densities that were present in the early universe.

The search techniques have to be innovative. After all, scientists are hunting material that, by definition, we can't see and, aside from exerting a gravitational force, that doesn't interact with other matter.

The laboratory methods for directly detecting and measuring dark matter are very difficult, but attempting to identify and understand dark matter is well worth the effort. As the dominant form of matter in the universe, dark matter profoundly influenced the past development of the universe and will affect its future as well.

Looking for WIMPs and other microscopic dark matter

Astronomers discovered dark matter through their observations of galaxies. Now physicists are conducting experiments to find dark matter particles and learn what they are.

Here are some dark matter experiments that are currently underway:

>> Large particle detectors lie in deep underground laboratories, where the surrounding rock reduces interference from cosmic rays (high-speed, electrified subatomic particles of known types that come from space). As Earth moves through the dark matter of the Milky Way, dark matter particles may be found when they strike the detectors.

>> Telescopes are equipped to detect celestial gamma rays, on spacecraft such as NASA's Fermi Gamma-ray Space Telescope. An unusual feature in the spectrum of gamma rays may indicate that some of them come from dark matter particles annihilating in space.

>> Arrays of telescopes on Earth detect flashes of visible light that are triggered when celestial gamma rays strike the atmosphere. A prime example is the High Energy Stereoscopic System (HESS) in Namibia. Analysis of these data, like the Fermi data, can reveal a spectral feature of the incoming gamma rays that is traceable to dark matter at their source.

>> The Alpha Magnetic Spectrometer (AMS-02) operates on the International Space Station. AMS is searching for unusual cosmic rays that may be pro-duced by *neutralinos,* a certain type of theorized dark matter particles. When neutralinos collide with each other in space, they may create the cosmic rays that AMS is looking for.

Advances in technology are making other experiments concerning dark matter possible:

>> Powerful atom smashers, such as the Large Hadron Collider (LHC), near Geneva, Switzerland, can make subatomic particles knock into each other at very high energies. They can generate dark matter particles for detection in the laboratory.

>> Underground neutrino observatories (see Chapter 10) can be upgraded to make improved measurements of neutrinos from the Sun, to reveal physical conditions near the solar center. These experiments will test the theory that the Sun has accumulated a great deal of dark matter, concentrated toward its center.

For more information on some of these dark matter experiments, consult the following sites:

>> **NASA's Fermi Gamma-ray Space Telescope website** (`www.nasa.gov/content/fermi-gamma-ray-space-telescope`): Find news of the dark matter search and other Fermi findings.

>> **The Alpha Magnetic Spectrometer website** (`ams.nasa.gov`): A counter tells you how many billion cosmic rays it has measured since it was installed on the International Space Station in May 2011.

>> **The High Energy Stereoscopic System website** (`www.mpi-hd.mpg.de/hfm/HESS`): Read up on the HESS's many discoveries about astronomical sources of gamma rays.

>> **Large Hadron Collider** (`home.cern/topics/large-hadron-collider`): This site is the public website of the LHC, operated by CERN, the European Organization for Nuclear Research.

MACHOs: Making a brighter image

Because MACHOs aren't microscopic like WIMPS, looking for them is easier. The prime method takes advantage of a mind-bending concept from Einstein's General Theory of Relativity. To wit: Mass distorts the fabric of space and the path of a light wave (as I describe in Chapter 11), which means that the mass of an object that happens to lie along the line of sight between Earth and a distant star focuses the light from that star, briefly making it appear brighter. The more massive the object — in this case, a MACHO — the brighter the star appears during the alignment.

In effect, the MACHO acts as a miniature gravitational lens, or microlens, bending and brightening the light from the background star. (See Chapter 11 for more on microlensing.)

To search for MACHOs, astronomers monitor simultaneously the brightness of vast numbers of stars in one of the Milky Way's nearest neighbors, the Large Magellanic Cloud (LMC) galaxy. To reach Earth, starlight from the LMC must pass through the halo of the Milky Way, and MACHOs that reside in the halo should have a measurable effect on that light.

The searches record occasional events in which one star or another in the LMC seems to brighten unexpectedly and then fade again, as though it was gravitationally lensed by the mass of a MACHO passing in front of it and located in the halo of the Milky Way. But the frequency with which these events occur is so small that there can't be enough MACHOS (whatever they are) to account for any significant part of the dark matter in the galactic halo. Presumably the same is true of the dark matter in halos of other galaxies.

Mapping dark matter with gravitational lensing

On much larger scales, scientists are taking advantage of gravitational lensing to map the dark matter in galaxies, clusters of galaxies, and even larger regions of the universe.

If a cluster happens to lie in the travel path of light from a background galaxy, it bends and distorts the light as I explain in the earlier section "Dark matter causes gravitational lensing," creating multiple images of the background body. A halo of these ghost images forms in and around the cluster, as seen from Earth. Astronomers have used the Hubble Space Telescope to photograph some clusters of galaxies where large numbers of the ghost images of a more distant galaxy appear as short, bright arcs seen against a cluster.

To create the exact pattern of observed ghost images, the intervening cluster must have its mass distributed in a particular way. Because most of the cluster's mass consists of dark matter, the process of gravitational lensing reveals how the dark matter is concentrated in the cluster.

Astronomers made an important test of the theory of cold dark matter (CDM) when they combined several different kinds of observations of the Bullet Cluster, and gravitational lensing was a key part of the test. "Bullet Cluster" (or "Bullet") is the nickname for a pair of clusters of galaxies, close together in the sky, that appear to have collided and gone through each other. When galaxies collide, most of the stars of one galaxy will zip right through the other galaxy and wind up on the other side because stars in a galaxy are very far apart from each other compared to their individual sizes. In other words, a galaxy is mostly empty space, despite the millions or billions of stars within it. The same is true when clusters of galaxies collide: The whole clusters pass through each other, mostly intact, and land on opposite sides from where they began. Except clusters of galaxies are often filled with very hot gas that glows in X-rays (as I mention in the earlier section "Dark matter makes stars orbit oddly"), and, of course, there's dark matter throughout clusters, as Zwicky first suspected.

A key part of the CDM theory is that although dark matter particles are affected by gravity, no other force affects them very much, if at all. Astronomers observed the

Bullet with several telescopes in space and on the ground and used them as follows:

>> Images made in X-rays showed where the hot gas from the two clusters ended up after the two passed through each other.

>> Images made in infrared and visible light showed where the clusters of galaxies themselves are located in the aftermath of the collision. They also showed many more distant galaxies, beyond the Bullet, which were gravitationally lensed by the dark matter in the Bullet.

From these images, the astronomers were able to deduce the following:

>> The two clusters passed through each other in the collision.

>> The hot gas that was originally *inside* each of the clusters is now *between* the two clusters.

>> The dark matter that was originally inside each cluster moved with the cluster and is still inside it.

The astronomers concluded that the clouds of hot gas from the two clusters slowed each other down by so-called ram pressure when the two clouds collided, so neither cloud stayed inside its own cluster, as the clusters raced ahead. Ram pressure is the force exerted on a body when it moves through a fluid (fluids can be liquid or gaseous; this fluid was hot intracluster gas). The dark matter of each moving cluster stayed with the cluster because, as the CDM theory requires, it's not subject to frictional or other forces such as ram pressure.

Dueling Antimatter: Proving That Opposites Attract

Get ready for another type of matter almost as weird as dark matter — or maybe even weirder. I'm talking about antimatter.

TECHNICAL STUFF

British physicist Paul Dirac predicted the existence of antimatter in 1929. He combined the theories of quantum mechanics, electromagnetism, and relativity in an elegant set of mathematical equations. (If you want to know more about his theories, you'll have to look them up; this isn't a physics book.)

Dirac found that, for every subatomic particle, a mirror-image twin should exist, identical in mass but with an opposite electrical charge. Thus, the proton has its antiproton and the electron its antielectron.

When a particle and its antiparticle meet, they annihilate each other. Their electric charges cancel out, and their mass is converted into pure energy.

Astronomers have detected antiparticles of the electron and proton in the cosmic rays from deep space. The antielectron is called the *positron*, and the antiproton is simply the *antiproton*. The Alpha Magnetic Spectrometer (AMS-02) on the International Space Station, which I mention earlier in this chapter, is searching for antihelium that also may exist in cosmic rays. As of December 2016, after five years in space, AMS-02 had detected more than 90 billion cosmic rays, but not a single one was definitely identified as antihelium. Physicists have actually made antiparticles and even entire antiatoms, such as antihydrogen, in the laboratory. Doctors use beams of antiparticles to diagnose and treat cancer.

Physicists had more success with antiprotons. They found a belt of antiprotons within Earth's Van Allen radiation belts (which I describe in Chapter 5). Researchers from Italy and elsewhere discovered the antiproton belt in 2011 with the PAMELA (Payload for Antimatter Matter Exploration and Light-nuclei Astrophysics) particle detector experiment on Russia's Resurs DK1 satellite.

Astronomers studying high-energy radiation from space regularly observe gamma rays that are known as *annihilation radiation.* When an electron and its antiparticle, the positron, meet, they annihilate, releasing gamma rays at a known energy of 511 kiloelectron volts (keV). These telltale rays have been detected from several places in the galaxy, including a wide region in the direction toward the center of the Milky Way. (You can see a map of the annihilation radiation from the Milky Way, as measured by the European Space Agency's INTEGRAL satellite, at `sci.esa.int/integral/45328-integral-maps-the-galaxy-at-511-kev`.) Annihilation radiation has also been detected from some powerful solar flares (see Chapter 10 for more about solar flares).

On the cosmic scale, the big mystery is why the universe contains many more particles than antiparticles. Experiments are underway to find the answer. Many physicists think that the Big Bang forged equal numbers of both. If that's what happened, an unknown physical process has since altered that original balance between matter and antimatter in favor of matter. On the other hand, recent studies suggest that the imbalance was always there and is a normal part of nature.

At least we know that we have billions of years to solve the problem of the relative lack of antimatter before the universe (and us with it!) slips away to its ultimate fate (which I discuss in the next chapter). But all in all, I'm pro antimatter.

Ron Cowen, who writes about astronomy and space for many publications, originally contributed this chapter. The author, Stephen P. Maran, updated it for the second and all subsequent editions of Astronomy For Dummies. *All opinions expressed in this chapter are those of the author.*

Chapter **16**

The Big Bang and the Evolution of the Universe

O nce upon a time, 13.8 billion years ago, the universe as we know it didn't exist. No matter, no atoms, no light, no photons — not even space or time.

Suddenly, perhaps in an instant, the universe took form as a tiny, dense speck filled with light. In a minuscule fraction of a second, all the matter and energy in the cosmos came into being. Much smaller than an atom, the infant universe was searingly hot, a fireball that began mushrooming in size and cooling at a furious rate.

Astronomers and people the world over have come to know this picture of the birth of the universe as the *Big Bang* theory.

The Big Bang wasn't like a bomb that explodes into the environment — there was no environment until the Big Bang occurred. It was the origin and rapid expansion of space itself. During the first trillion-trillion-trillionth of a second, the universe grew more than a trillion-trillion-trillion times bigger. From an original smooth mixture of subatomic particles and radiation arose the collection of galaxies, galaxy clusters, and superclusters present in the universe today. It boggles the mind to think that the largest structures in the universe, congregations of galaxies that stretch hundreds of millions of light-years across the sky, began as subatomic fluctuations in the energy of the infant cosmos. But that's what scientists believe about how the universe took shape.

In this chapter, I cover evidence supporting the Big Bang theory, the expansion of the universe, and related information on dark energy, the cosmic microwave background, the Hubble constant, and standard candles.

TIP

For more information on the concepts in this chapter, visit UCLA's Frequently Asked Questions in Cosmology site, at www.astro.ucla.edu/~wright/cosmology_faq.html. The site is maintained by Professor Ned Wright, who gives you the right stuff.

Evidence for the Big Bang

Why believe that the universe began with a bang?

Astronomers cite three different discoveries that make a compelling case for the theory:

>> **The expanding universe:** Perhaps the most convincing evidence for the Big Bang comes from a remarkable discovery made by Edwin Hubble in 1929. Up to that time, most scientists viewed the universe as *static,* meaning never changing in its entirety. But Hubble discovered that the universe is expanding. Groups of galaxies are flying away from each other, like debris flung in all directions from a cosmic explosion, but they haven't been flung apart into space; the space itself between them is expanding, which makes them move farther and farther apart.

Astronomer and Catholic priest Georges Lemaitre was the first to reason that if galaxies are flying apart, they were once closer together. Tracing the expansion of the universe back in time, astronomers — aided by telescopes and by observatories in space — found that, 13.8 billion years ago (give or take 100 million years), the universe was an incredibly hot, dense place in which a tremendous release of energy triggered an enormous explosion.

>> **The cosmic microwave background:** In the 1940s, physicist George Gamow realized that a Big Bang would produce intense radiation. His colleagues suggested that remnants of this radiation, cooled by the expansion of the

universe, may still exist — like the fumes that persist from an extinguished house fire.

In 1964, Arno Penzias and Robert Wilson of Bell Laboratories were scanning the sky with a radio receiver when they detected a faint, uniform crackling. What the researchers first assumed was static in their receiver turned out to be the faint whisper of radiation left over from the Big Bang. The radiation is a uniform glow of microwave radiation (short radio waves) permeating space. This *cosmic microwave background* has exactly the temperature that astronomers calculate it should (2.73 K above absolute zero, if it has cooled steadily since the Big Bang. (Absolute zero is –273.15°C or –459.67°F.) For their historic discovery, Penzias and Wilson shared the 1978 Nobel Prize in physics. (See the section "Universal Info Pulled from the Cosmic Microwave Background" later in this chapter for the full scoop.)

>> **The cosmic abundance of helium:** Astronomers have found that the amount of helium among all the baryonic matter in the universe is 24 percent by mass (the rest of baryonic matter is almost entirely hydrogen; iron, carbon, oxygen, and all that good stuff put together is just a trace constituent, compared to hydrogen and helium). Nuclear reactions inside stars (see Chapter 11) haven't gone on long enough to produce this amount of helium. But the helium we've detected is just the amount that the theory predicts would've been forged in the Big Bang. Adding to the evidence, NASA's Wilkinson Microwave Anisotropy Probe (WMAP) found that there was helium in the early universe before there were stars.

Astronomers have other observational evidence that the universe has expanded and changed with time besides these three discoveries. For example, the deepest photographs of space made with the Hubble Space Telescope reveal that galaxies in the early universe were often smaller or more irregular in shape and more likely to be colliding with each other than present-day galaxies. That information is consistent with the idea that the universe was much smaller back then, so galaxies were closer together and more likely to collide. If the universe had recently formed from a Big Bang, the galaxies themselves would have been younger and smaller.

Such evidence makes clear that the universe is evolving in ways that are consistent with the idea that it began with the Big Bang and has gotten bigger over time.

As successful as the standard Big Bang theory has proved to be in accounting for observations of the cosmos, the theory is but a starting point for exploring the early universe. For example, despite its name, the theory doesn't suggest a source for the cosmic dynamite that sparked the Big Bang in the first place.

Inflation: A Swell Time in the Universe

Aside from ignoring the source of the expansion-causing explosion, the Big Bang theory has other shortcomings. In particular, it doesn't explain why regions of the universe that are separated by distances so vast that they can't communicate — even by a messenger traveling at the speed of light — look so similar to each other.

In 1980, physicist Alan Guth devised a theory, which he called *inflation*, that can help explain this puzzle. He suggested that a tiny fraction of a second after the Big Bang, the universe underwent a tremendous growth spurt. In just 10^{-32} seconds (a hundred-millionth of a trillionth of a trillionth of a second), the universe expanded at a rate far greater than at any time in the 13.8 billion years that have elapsed since.

This period of enormous expansion spread tiny regions — which had once been in close contact — to the far corners of the universe. As a result, the cosmos looks the same on the large scale, no matter what direction you point a telescope. (Think of a big, lumpy ball of dough: If you roll the dough over and over again with your rolling pin, eventually you smooth out all the lumps and create a uniform sheet of dough.) Indeed, inflation expanded tiny regions of space into volumes far bigger than astronomers can ever observe. This expansion suggests the intriguing possibility that inflation created universes far beyond the scope of our own. Instead of a single universe, a collection of universes, or a *multiverse*, may exist. But I'm adverse to that theory — one universe is hard enough to comprehend as it is!

Inflation had another effect: The infinitesimally short but extraordinarily great growth spurt after the Big Bang captured random, subatomic fluctuations in energy and blew them up to macroscopic proportions. By preserving and amplifying these so-called *quantum fluctuations*, inflation produced regions of the universe with slight variations in density from one to another.

Because of inflation and the quantum fluctuations, some regions of the universe contain more matter and energy, on average, than other regions. As a result, there are cold spots and hot spots in the temperature of the cosmic microwave background (see Figure 16-1). Over time, gravity molded these variations into the spidery networks of galaxy clusters and giant voids that fill our universe today, as I describe in Chapter 12. Check out "Universal Info Pulled from the Cosmic Microwave Background" later in this chapter for more information.

The following sections cover two other interesting facets of inflation: the vacuum where inflation gets its power and the relationship between inflation and the universe's shape.

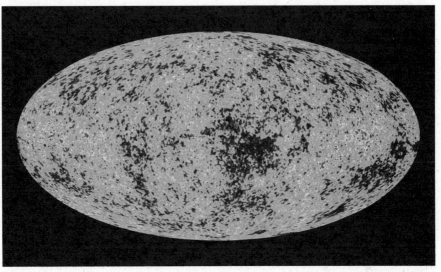

FIGURE 16-1:
A "baby picture"
of the universe,
from the
Wilkinson
Microwave
Anisotropy Probe
satellite.

Something from nothing: Inflation and the vacuum

Ironically, the reservoir of energy that powers inflation comes from nothing: the *vacuum*. According to quantum theory, the vacuum of space is far from empty. It seethes with particles and antiparticles that are constantly being created and destroyed. Tapping into this energy, theorists suggest, provided the Big Bang with its explosive energy and radiation.

The vacuum has another bizarre property: It can exert a repulsive force. Instead of pulling two objects together as gravity does, this force pushes them farther apart. The repulsive force of the vacuum may have led to the brief but powerful era of inflation.

Just as in the economy, cosmic inflation generates high interest. And this bubble won't burst.

Falling flat: Inflation and the shape of the universe

The inflation process — at least, in its simplest imagined form — would have imposed another condition on the universe: making the geometry of the universe flat. This rapid period of expansion would have stretched out any curvature in the cosmos like a balloon blown to enormous proportions.

For the universe to be flat, it must contain a very specific density, called the *critical density*. If the density of the universe is greater than the critical value, gravity's pull will be strong enough to reverse the expansion, eventually causing the universe to collapse into what astronomers call the *Big Crunch*.

Such a universe would curve back on itself to form a closed space of finite volume, like the surface of a sphere. A starship traveling in a straight line would eventually find itself back where it started. Mathematicians call this geometry *positive curvature*.

If the density is less than the critical value, gravity can never overpower the expansion, and the universe will continue to grow forever. Such a universe has *negative curvature*, with a shape akin to a horse's saddle.

Although inflation theory demands that the universe is flat, several types of observations have revealed that the universe doesn't have enough matter (whether normal or dark matter — see Chapter 15) to attain critical density.

So if the universe is flat, matter as we know it — or even as we don't know it — can't do the trick. But like Mighty Mouse, energy can save the day! In fact, it can save the universe, and recent research shows that it does. The data contained in the "baby picture of the universe" shown in Figure 16-1, which is a sky map of the cosmic microwave background radiation as measured by the WMAP satellite, has convinced essentially all cosmologists that the universe is flat and that energy is responsible. But it's not energy as we ever knew it before; dark energy is the hero. Read on to discover the dark side.

Dark Energy: The Universal Accelerator

Dark energy has a startling effect: It exerts a repulsive force throughout the universe. That's all scientists know about it. We don't know what dark energy is, so we define it by its observable property, the repulsive force. After the Big Bang and inflation occurred, gravity slowed the expansion of the universe. But as the universe grew bigger and bigger so the matter was spread apart into more and more space, the slowing effect of gravity became weaker. After a while (some billions of years), dark energy's repulsive force took over, causing the universe to expand ever faster. Observations from the Hubble and other telescopes have revealed this bizarre phenomenon.

The observations that revealed the existence of dark energy — by showing that the expansion of the universe is speeding up — were of Type Ia supernovas in distant galaxies. (You can read about Type Ia and other supernovas in Chapter 11.) All supernovas are bright enough to be seen in distant galaxies, but the Ia supernovas have a special property. Astronomers believe that these explosions all have roughly the same intrinsic brightness, like incandescent light bulbs of a particular wattage (see the section "In a Galaxy Far Away: Standard Candles and the Hubble Constant" later in this chapter).

Because light from a distant galaxy takes hundreds of millions of years or more to reach Earth, observations of that galaxy may show supernovas that erupted when the universe was much younger. If the expansion of the universe had been slowing ever since the Big Bang, there would be less distance between Earth and the far-away galaxy — and a shorter travel time for light — than if the universe had continued to expand at a fixed speed. Thus, in the case of a slower expansion, a supernova from a distant galaxy should look slightly brighter.

But in 1998, two teams of astronomers found exactly the opposite result: Distant supernovas looked slightly dimmer than expected, as if their home galaxies were farther away than calculated. It appears that the universe has revved up its rate of expansion. This discovery revealed the presence of dark energy, as I describe in Chapter 11, and earned the Nobel Prize in Physics for three astronomers — Saul Perlmutter, Adam Riess, and Brian Schmidt — who led the research.

Universal Info Pulled from the Cosmic Microwave Background

The cosmic microwave background (the faint whisper of radiation left over from the Big Bang) represents a snapshot of the universe when it was 379,000 years old. Before that time, a fog of electrons pervaded the infant universe, and radiation created in the Big Bang couldn't stream freely through space. The negatively charged particles repeatedly absorbed and scattered the radiation.

Around the time that the cosmos celebrated its 379,000th birthday, the universe became cool enough for electrons to combine with atomic nuclei, meaning there wasn't an abundance of particles to scatter and absorb radiation. The absorbing fog was lifted. Today we detect the light from the universe at age 379,000 years — now shifted in wavelength by the expansion of the universe — as microwaves and far-infrared light.

Finding the lumps in the cosmic microwave background

When Penzias and Wilson first detected the cosmic microwave background in the 1960s, it appeared to have a perfectly uniform temperature across the sky. No regions in the sky were ever so slightly hotter or colder — at least, not to the detection limits of the available instruments. That uniformity was a puzzle because such tiny variations in temperature have to be present to explain how the universe could have begun as a smooth soup of particles and radiation and evolved into a lumpy collection of galaxies, stars, and planets.

According to theory, the infant universe wasn't perfectly smooth. Like lumps in a bowl of porridge, it had slightly overdense and underdense places, with more atoms per cubic inch or fewer atoms per cubic inch, respectively. These places represent the tiny seeds around which matter could have started to clump together to form galaxies. Scientists should now see the variations in density as tiny fluctuations or anisotropies in the temperature of the cosmic microwave background. (An *anisotropy* is a difference in the physical properties of space, such as temperature and density, along one direction from the properties in another direction.)

In 1992, NASA's Cosmic Background Explorer (COBE) satellite, which just three years earlier had measured the temperature of the microwave background to an unprecedented accuracy, achieved what many astronomers consider an even greater triumph: It detected hot and cold spots in the cosmic microwave background. The COBE measurements earned the Nobel Prize in Physics for 2006 for both my NASA colleague John Mather and George Smoot of the University of California, Berkeley. (I never came close to a Nobel Prize [nor deserved one], but for years John Mather's office was just a few steps down the hall from mine. It's a small universe.)

The variations are indeed minuscule — less than 10,000th of a degree K colder or hotter than the average temperature of 2.73 K. The princess who could feel a pea through a big stack of mattresses wouldn't have felt these differences. Nonetheless, these cosmic ripples are large enough to account for the growth of structure in the universe. You can sleep on that.

Mapping the universe with the cosmic microwave background

In the search to find out whether the universe is flat or saddle-shaped, scientists looked to the cosmic microwave background for answers. A flat universe would dictate that the temperature fluctuations have a particular pattern. A slew of balloon-borne and ground-based telescopes suggested that the microwave background may have this pattern.

In 2003, NASA reported that its Wilkinson Microwave Anisotropy Probe had mapped and measured the microwave background over the entire sky in sharper detail than ever. The WMAP team, led by Charles Bennett, answered most of the existing questions about the Big Bang except what made it happen and what exactly is dark energy. In particular, they found that the universe is flat (a condition I define earlier in "Falling flat: Inflation and the shape of the universe"). That's consistent with the theory of inflation at the beginning of the Big Bang, which I also explain earlier.

Another satellite, the European Space Agency's Planck, operated from 2009 to 2013, making similar measurements to WMAP (and some of them more precise). Besides confirming inflation, these two satellites taught us the following:

» The present age of the universe is 13.8 billion years, a number I quote throughout this book.

» The cosmic microwave background radiation originated when the universe was 379,000 years old.

» The first stars began to shine about 200 million years after the Big Bang.

» The universe is flat, consistent with the theory of inflation (see the section "Inflation: A Swell Time in the Universe" earlier in this chapter).

» The relative amounts of mass energy in the universe are as follows:

- Normal matter (baryonic matter like that found on Earth): 4.9 percent

- Dark matter (see Chapter 15): 26.8 percent

- Dark energy: 68.3 percent

Scientists had made rough estimates for all these quantities, but now they have precise values. Even so, some experts give equally precise, but slightly different values for these cosmic numbers.

TIP

You can read all about WMAP and its findings on the Probe's official website at Goddard Space Flight Center, map.gsfc.nasa.gov. Check out the animations of the evolution of the universe and other cosmic topics on the site.

In a Galaxy Far Away: Standard Candles and the Hubble Constant

One of the longest-running questions in astronomy used to be "How old is the universe?" Now, thanks to WMAP, the Hubble Space Telescope, and other instruments, we know that the answer is 13.8 billion years.

So how did scientists figure out this magic number? They relied on information connected to the expansion of the universe: standard candles, which astronomers use to measure the distances of galaxies; and the Hubble constant, which relates galaxy distances to the rate at which the universe is expanding. I cover these topics in the following sections.

Standard candles: How do scientists measure galaxy distances?

Most strategies for measuring distance require some kind of *standard candle*, the cosmic equivalent of a light bulb of known wattage.

For instance, suppose that you believe you know the true brightness, or *luminosity*, of a particular type of star. Light from a distant source grows dimmer in proportion to the square of the distance, so the apparent brightness of a star of that same type in a distant galaxy indicates how far away the galaxy lies.

Yellowish, pulsating stars known as *Cepheid variables* remain one of the most credible standard candles for estimating the distance to relatively nearby galaxies (see Chapter 12). These youthful stars brighten and dim periodically. In 1912, Henrietta Leavitt of Harvard College Observatory detected that the rapidity with which Cepheids change their brightness is directly linked to their true luminosity. The longer the period, the greater the luminosity. A century has gone by since Leavitt's discovery, and astronomers still use Cepheids to measure distances in space.

Type Ia supernovas (see Chapter 11) are another type of standard candle. Because supernovas are much brighter than Cepheids, we can observe them in much more distant galaxies. Recent calculations of the Hubble constant employed both of these candles and got results that were in good agreement with each other and with the data from the WMAP satellite.

The Hubble constant: How fast do galaxies really move?

Cosmic age estimates have depended on a number that has held the attention of astronomers for decades: the *Hubble constant,* which represents the rate at which the universe is currently expanding. The number was named for Edwin Hubble, who found that we live in an expanding universe. In particular, he made the remarkable discovery that every distant galaxy (those beyond the Local Group of Galaxies, which I describe in Chapter 12) appears to be racing away from our home galaxy, the Milky Way.

Hubble found that the more remote the galaxy, the faster it recedes. This relationship is known as *Hubble's Law.* For example, consider two galaxies, one of which lies twice as far from the Milky Way as the other. The galaxy that resides twice as far away appears to move away twice as fast. (According to Albert Einstein's General Theory of Relativity, the galaxies themselves don't move; instead, the fabric of space in which they reside expands.)

TECHNICAL
STUFF

The constant of proportionality that relates the distance of a galaxy to its recession speed is known as the Hubble constant, or H_o. In other words, the speed at which a galaxy recedes is equal to H_o multiplied by the galaxy's distance. H_o thus provides a measure of the rate of universal expansion and, by implication, its age. (If you know how far away a galaxy is now and the rate at which it has been moving, you can calculate how long it took to get that far. According to the Big Bang theory, the universe was once infinitesimal in size, and then space began to expand. The points in space where we are now and where a particular galaxy is now were once right on top of each other, but as the universe aged, the points moved apart. The time they took to move to their present distance from each other is the age of the universe.)

TECHNICAL
STUFF

The Hubble constant is measured in kilometers per second per megaparsec. (One megaparsec is 3.26 million light-years.) At one time, top experts disagreed in their measurements of the Hubble constant by nearly a factor of two. Then, after years of study, astronomers using the Hubble Space Telescope reported a value of 70 for the Hubble constant. That number means that a galaxy about 30 megaparsecs (about 100 million light-years) from Earth speeds away at 2,100 kilometers per second, which is about 1,300 miles per second. However, recent observations disagree with each other by up to 8 percent. The correct Hubble constant is somewhere between 67 and 73. Or is it? Some astronomers think the value of the constant may be different at different places in space. I think we should be happy that all the measurements agree to within less than 10 percent.

Thanks to standard candles and the Hubble constant, astronomers now have reliable data on the current expansion rate of the universe, and we know that dark energy is increasing the expansion rate. But the nature of dark energy remains a deep, dark mystery.

The Fate of the Universe

Dark energy makes the universe expand faster and faster as time goes on. The Hubble constant thus doesn't stay constant very long; it gets bigger. In other words, the Hubble constant is more of a "Hubble inconstant."

As the universe keeps expanding at greater and greater rates, eventually other galaxies will be moving away from us at a rate greater than the speed of light. When you read the preceding sentence, you may have said, "Wait a minute! In Chapter 13, you wrote that things can't go faster than the speed of light, except for tachyon particles, which may not even exist. So what's the deal with these galaxies?"

The answer is that, one day trillions of years in the future, the galaxies that will be moving away at a rate greater than light speed won't be doing that moving themselves. At the beginning of this chapter, I told you that the Big Bang "was the origin and rapid expansion of space itself." The seemingly high speeds of the galaxies that are rushing apart aren't actual motions of galaxies; they're caused by the expansion of space itself. Space isn't matter, and it can go as fast as dark energy makes it expand.

When other galaxies are moving away at above the speed of light, their light will no longer reach our Milky Way galaxy. The Sun will be gone long before that — it will have used up its core hydrogen 4 billion years from now (see Chapter 11 for details) — and, soon afterward, will become a red giant, shed its outer layers, and fade away as a white dwarf. But there may be other stars in the Milky Way, still going strong, with planets and maybe even intelligent beings. Those extraterrestrials won't see the galaxies whose light can't reach them. The universe outside of their galaxy, in effect, will go dark.

At one time, astronomers thought the universe would remain much as we know it far into the future. But the discovery of dark energy changed all that. As Yogi Berra famously said, "The future ain't what it used to be."

Ron Cowen, who covers astronomy and space for many publications, originally contributed this chapter. The author, Stephen P. Maran, updated it for the subsequent and current editions of Astronomy For Dummies. *All opinions expressed in this chapter are those of the author.*

5 The Part of Tens

IN THIS PART . . .

Discover ten odd facts about space that you can use to impress your friends.

Get the lowdown on ten mistakes that people and the media frequently make on the topic of astronomy.

and the Big Bang on television

» Finding out why Pluto's discovery was an accident, why sunspots aren't dark, and why rain never hits the ground on Venus

» Exploring tidal myths, exploding stars, and Earth's uniqueness

Chapter 17

Ten Strange Facts about Astronomy and Space

Here are some of my favorite facts about astronomy and, in particular, Earth and its solar system. With the following information under your belt, you may be ready to handle the astronomy questions on television quiz shows and inquiries from friends and family.

You Have Tiny Meteorites in Your Hair

Micrometeorites, tiny particles from space visible only through microscopes, are constantly raining down on Earth. Some fall on you whenever you go outdoors. But without the most advanced laboratory equipment and analysis techniques, you can't detect them. They get lost in the great mass of pollen, smog particles, household dust, and (I'm sorry to say) dandruff that resides on the top of your head. (Check out Chapter 4 for the scoop on meteorites of all sizes.)

A Comet's Tail Often Leads the Way

A comet tail isn't like a horse tail, which always trails behind as the horse gallops ahead. A comet tail always points away from the Sun. When a comet approaches the Sun, its tail, or tails, stream behind it; when the comet heads back out into the solar system, the tail leads the way. (See Chapter 4 for more information about comets.)

Earth Is Made of Rare and Unusual Matter

The great majority of all the matter in the universe is so-called *dark matter*, invisible stuff that astronomers haven't yet identified (see Chapter 15). And most ordinary or visible matter is in the form of plasma (hot, electrified gas that makes up normal stars such as the Sun) or degenerate matter (in which atoms or even the nuclei within the atoms are crushed together to unimaginable density, as found in white dwarfs and neutron stars; see Chapter 11). You don't find dark matter, degenerate matter, or much plasma on Earth. Compared to the great bulk of the universe, Earth and earthlings are the aliens. (See Chapter 5 for more about Earth's unique properties.)

High Tide Comes on Both Sides of Earth at the Same Time

Ocean tides on the side of Earth that faces the Moon aren't appreciably higher than tides on the opposite side of Earth at the same time. This may defy common sense, but not physics and mathematical analysis. (The same goes for the smaller ocean tides raised by the Sun.) See Chapter 5 for more about the Moon.

On Venus, the Rain Never Falls on the Plain

In fact, the constant rain on Venus never falls on anything. It evaporates before it hits the ground, and the rain is pure acid. (The common name for evaporating rain is *virga*; see Chapter 6 for more about Venus.)

Rocks from Mars Dot Earth

People have found about 100 meteorites on Earth that come from the crust of Mars, blasted from that planet by the impacts of much larger objects — perhaps from the asteroid belt (see Chapters 4 and 7 for info on meteorites and asteroids, respectively). Statistically, many more undiscovered Mars rocks must have fallen into the ocean or landed in out-of-the-way places where they haven't been spotted. (See Chapter 6 to find out more about Mars.)

Pluto Was Discovered from the Predictions of a False Theory

Percival Lowell predicted the existence and approximate location of the object that we now call Pluto. When Clyde Tombaugh surveyed the designated region, he discovered Pluto. But now scientists know that Lowell's theory, which inferred the existence of Pluto from its gravitational effects on the motion of Uranus, was wrong. In fact, Pluto's mass is much too small to produce the "observed" effects. Furthermore, the "gravitational effects" were just errors in measuring the motion of Uranus. (Not enough information was available about Neptune's motion to study it for clues.) The discovery of Pluto took hard work, but as it happened, it was just plain luck. And although Lowell predicted the existence of a planet, as Pluto was first termed, the International Astronomical Union has since downgraded it to dwarf planet. (See Chapter 9 to find out all about Pluto.)

Sunspots Aren't Dark

Almost everyone "knows" that sunspots are "dark" spots on the Sun. But in reality, sunspots are simply places where the hot solar gas is slightly cooler than its surroundings (see Chapter 10 for more explanation). The spots look dark compared to their hotter surroundings, but if all you can see is the sunspot, it looks bright.

A Star in Plain View May Have Exploded, but No One Knows

Eta Carinae is one of the most massive, fiercely shining stars in our galaxy, and astronomers expect it to produce a powerful supernova explosion at any time, if it hasn't already. But because light takes about 8,000 years to travel from Eta Carinae to Earth, an explosion that occurred less than that many years ago isn't visible to us yet. (See Chapter 11 to discover more about the life cycles of stars.)

You May Have Seen the Big Bang on an Old Television

The Big Bang Theory premiered in 2007, but the real Big Bang may have made its TV debut even before that. Some of the *snow* — a pattern of interference that looks like little white spots or streaks on old black-and-white television sets — was actually radio waves the TV antenna received from the cosmic microwave background, a glow from the early universe in the aftermath of the Big Bang (see Chapter 16). When this radiation was actually discovered at the Bell Telephone Laboratories, scientists studied many possible causes of the unexpected "noise" in the radio receiver. They even investigated pigeon droppings, or "white dielectric material" in science speak, as a possible cause but later dropped that suggestion.

Chapter **18**

Ten Common Errors about Astronomy and Space

n daily life — reading the newspaper, watching the evening news, surfing the web, catching up with social media, or talking to friends — you run across many misconceptions about astronomy. In this chapter, I explain the most common of these errors.

"The Light from That Star Took 1,000 Light-Years to Reach Earth"

Many people mistake the light-year for a unit of time on par with units like a day, a month, or an ordinary year. But a light-year is a unit of distance, equal to the length that light travels in a vacuum over a period of one year. (See Chapter 1.)

A Freshly Fallen Meteorite Is Still Hot

Actually, freshly fallen meteorites are cold; an icy frost (from contact with moisture in the air) sometimes forms on a frigid stone that has recently landed. When an eyewitness says that he saw a meteorite fall to the ground and that he burned his fingers on the rock, the account may be a hoax. (See Chapter 4 for more information about meteorites.) A quick trip through Earth's atmosphere isn't enough to substantially heat a rock that has spent the last several million years in the deep freeze of outer space.

Summer Always Comes When Earth Is Closest to the Sun

The belief that summer comes when Earth is closest to the Sun is about the most common error of them all, but common sense should tell you that the belief is false. After all, winter occurs in Australia when the United States is experiencing summer. But on any given day, Australia is just as far from the Sun as the United States. In fact, Earth is closest to the Sun in January and farthest from the Sun in July. (See Chapter 5 for more explanation of how the tilt of Earth's axis causes our seasons.)

The Back of the Moon Is Dark

Some people think that the back of the Moon, which faces away from Earth (astronomers call it the "far side") is dark. They may even call it the "dark side" of the Moon. Sometimes the far side *is* dark, but sometimes it is bright, and much of the time, part of it is dark and the rest of it is bright. In that way, the far side of the Moon is just like the near side, which faces Earth. When we see a full Moon, the near side is all bright and the far side is all dark. At new Moon, when the side of the Moon that faces Earth is dark, the far side of the Moon is bright — if we could see the far side then, it would look like a full Moon. For more about the Moon and its phases, check out Chapter 5.

The "Morning Star" Is a Star

The "morning star" isn't a star; it's always a planet. And sometimes two morning stars appear at once, such as Mercury and Venus (see Chapter 6). The same idea applies to the "evening star": You're seeing a planet, and you may see more than one. "Shooting stars" and "falling stars" are misnomers, too. These "stars" are meteors — the flashes of light caused by small meteoroids falling through Earth's atmosphere (see Chapter 4). Many of the "superstars" you see on television may be just flashes in the pan, but they at least get 15 minutes of fame.

If You Vacation in the Asteroid Belt, You'll See Asteroids All Around You

In just about any movie about space travel, you see a scene in which the intrepid pilot skillfully steers the spaceship past hundreds of asteroids that hurtle past in every direction, sometimes coming five at a time. Moviemakers just don't understand the vastness of the solar system, or they ignore it for dramatic purposes. If you stood on an asteroid smack dab in the middle of the main asteroid belt between Mars and Jupiter, you'd be lucky to see more than one or two other asteroids, if any, with the naked eye. (See Chapter 7 for more information about asteroids.) Even in the asteroid belt, outer space is mostly empty space.

Nuking a "Killer Asteroid" on a Collision Course for Earth Will Save Us

You come across many common errors about asteroids, and various Hollywood doomsday movies and media reports on "killer asteroids" have provided ample but unfortunate opportunities to reinforce these misunderstandings among the public.

Blowing up an asteroid on a collision course with Earth with an H-bomb may only create smaller and collectively just-as-dangerous rocks, all still heading for our planet. A less risky idea is to attach a rocket motor to gently propel the asteroid just the slightest bit forward or backward in its orbit, steering it so it doesn't get to the same place in space as Earth at the same time. Even better, we might launch a so-called gravity tractor satellite to gently pull the asteroid from its original trajectory so it misses Earth. (I explain the gravity tractor method in Chapter 7.)

The Sun Is an Average Star

You often hear or read statements that the Sun is an average star. In fact, the vast majority of all stars are smaller, dimmer, cooler, and less massive than our Sun (see Chapter 10). Be proud of the Sun — it's like a kid from the mythical Lake Wobegon, where the children are all "above average."

The Hubble Telescope Gets Up Close and Personal

The Hubble Space Telescope doesn't snap those beautiful pictures by cruising through space until it floats alongside nebulae, star clusters, and galaxies (see Chapter 12). The telescope stays in close orbit around Earth and just takes great photos. It does so because it has incredibly well-made optics and orbits far above the parts of Earth's atmosphere that blur our view with telescopes on the ground.

The Big Bang Is Dead

When an astronomer reports a finding that doesn't fit the current understanding of cosmology, members of the media are prone to pronouncing, "The Big Bang is dead." (See Chapter 16 for an explanation of the Big Bang.) But astronomers are simply finding differences between the observed expansion of the universe and specific mathematical descriptions of it. The competing theories — including one that fits the newly reported data — are consistent with the Big Bang; they just differ in the details. Just like Mark Twain's falsely reported 1897 demise, news accounts of the Big Bang's death are greatly exaggerated.

6

Appendixes

Use the star maps to help you locate constellations in the night sky.

Flip to the glossary when you encounter an astronomy term that you're not familiar with.

Appendix A

Star Maps

The following pages contain eight star maps — four of the Northern Hemisphere and four of the Southern Hemisphere — to start you on your starry way.

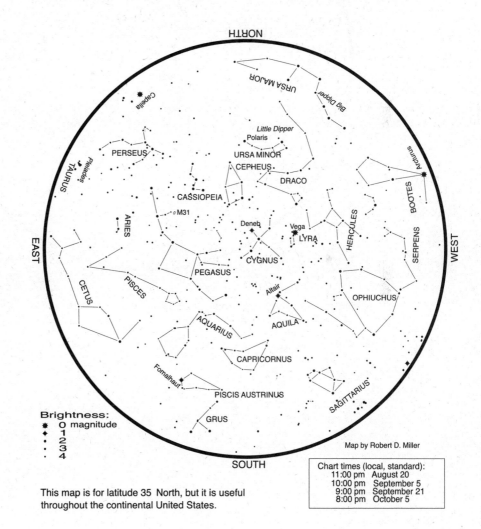

Brightness:
- 0 magnitude
- 1
- 2
- 3
- 4

Map by Robert D. Miller

This map is for latitude 35 North, but it is useful throughout the continental United States.

Chart times (local, standard):
- 11:00 pm   August 20
- 10:00 pm   September 5
- 9:00 pm   September 21
- 8:00 pm   October 5

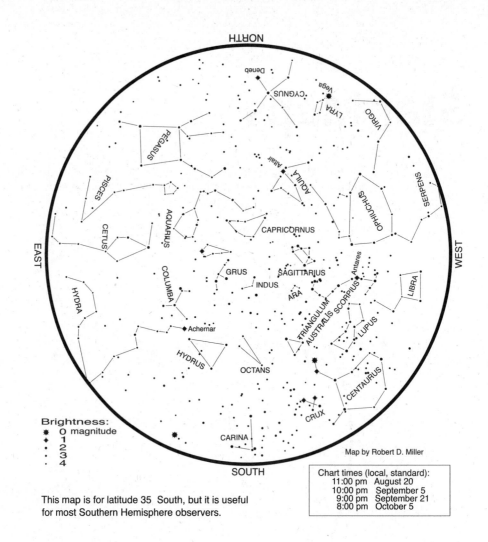

Brightness:
- ✱ 0 magnitude
- ◆ 1
- • 2
- • 3
- · 4

Map by Robert D. Miller

This map is for latitude 35 South, but it is useful for most Southern Hemisphere observers.

Chart times (local, standard):
11:00 pm	August 20
10:00 pm	September 5
9:00 pm	September 21
8:00 pm	October 5

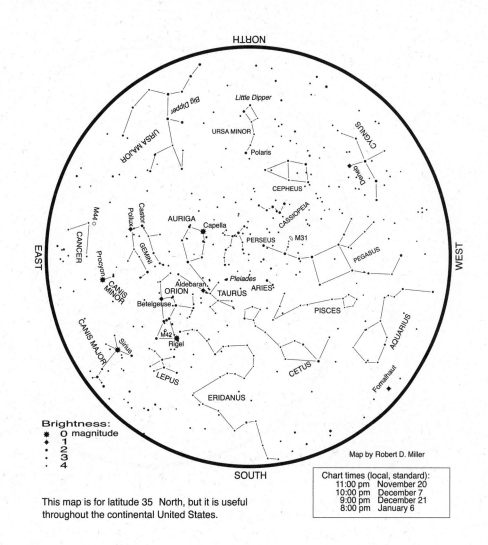

NORTH

Little Dipper

URSA MINOR

Polaris

CEPHEUS

CYGNUS

Deneb

URSA MAJOR

Big Dipper

CASSIOPEIA

M44

CANCER

Castor

Pollux

AURIGA

Capella

PERSEUS

M31

PEGASUS

Procyon

GEMINI

CANIS MINOR

Aldebaran

ORION

TAURUS

Pleiades

ARIES

PISCES

AQUARIUS

Betelgeuse

CANIS MAJOR

Sirius

M42

Rigel

LEPUS

CETUS

Fomalhaut

ERIDANUS

EAST

WEST

Brightness:

- ✳ 0 magnitude
- ◆ 1
- • 2
- · 3
- · 4

Map by Robert D. Miller

SOUTH

This map is for latitude 35 North, but it is useful throughout the continental United States.

Chart times (local, standard):
11:00 pm   November 20
10:00 pm   December 7
9:00 pm   December 21
8:00 pm   January 6

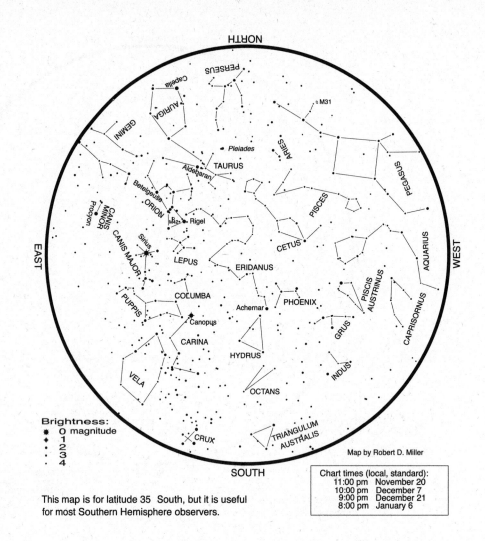

Brightness:
* 0 magnitude
* 1
* 2
* 3
* 4

NORTH

PERSEUS
Capella
AURIGA
GEMINI
Pleiades
TAURUS
Aldebaran
M31
ARIES
PISCES
PEGASUS
Betelgeuse
ORION
CANIS MINOR
Procyon
CANIS MAJOR
M42 Rigel
Sirius
CETUS
LEPUS
AQUARIUS
ERIDANUS
WEST
COLUMBA
PUPPIS
PHOENIX
PISCIS AUSTRINUS
CAPRISORNUS
Achernar
Canopus
CARINA
GRUS
HYDRUS
INDUS
VELA
OCTANS
TRIANGULUM AUSTRALIS
CRUX

EAST

Map by Robert D. Miller

SOUTH

This map is for latitude 35 South, but it is useful for most Southern Hemisphere observers.

Chart times (local, standard):
11:00 pm   November 20
10:00 pm   December 7
9:00 pm   December 21
8:00 pm   January 6

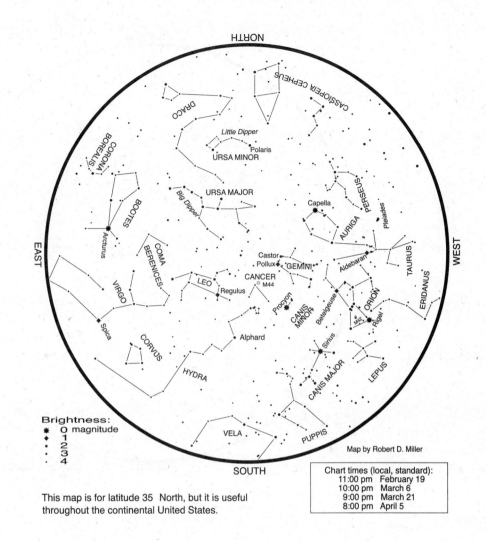

Brightness:
✴ 0 magnitude
◆ 1
• 2
· 3
· 4

This map is for latitude 35 North, but it is useful throughout the continental United States.

Map by Robert D. Miller

Chart times (local, standard):
11:00 pm   February 19
10:00 pm   March 6
9:00 pm   March 21
8:00 pm   April 5

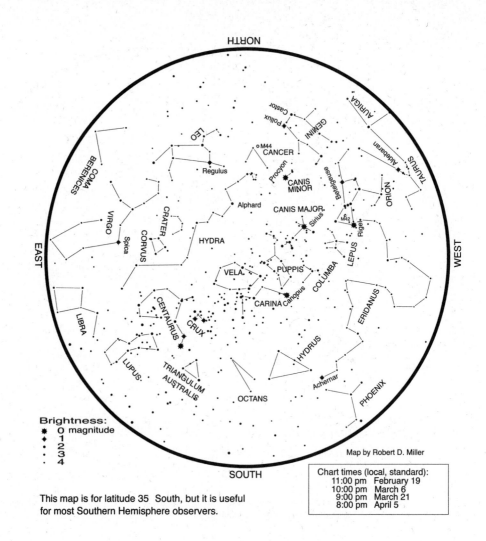

NORTH

GEMINI
Castor
Pollux
AURIGA
TAURUS
Aldebaran
ORION
Betelgeuse
CANIS MINOR
Procyon
CANCER
M44
LEO
Regulus
COMA BERENICES
Alphard
CANIS MAJOR
Sirius
M42
Rigel
LEPUS
COLUMBA
ERIDANUS
VIRGO
CRATER
HYDRA
PUPPIS
Spica
CORVUS
VELA
CARINA
Canopus
HYDRUS
EAST
WEST
LIBRA
CENTAURUS
CRUX
Achernar
PHOENIX
LUPUS
TRIANGULUM AUSTRALIS
OCTANS

Brightness:
✷ 0 magnitude
◆ 1
● 2
• 3
· 4

Map by Robert D. Miller

SOUTH

This map is for latitude 35 South, but it is useful
for most Southern Hemisphere observers.

Chart times (local, standard):
11:00 pm   February 19
10:00 pm   March 6
9:00 pm   March 21
8:00 pm   April 5

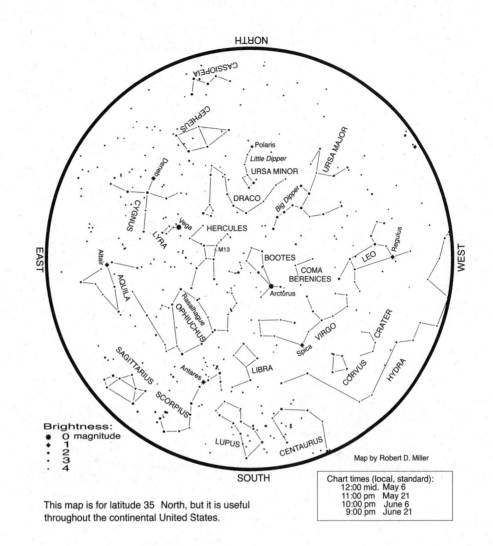

Brightness:
- ✴ 0 magnitude
- ◆ 1
- • 2
- · 3
- · 4

This map is for latitude 35 North, but it is useful throughout the continental United States.

Map by Robert D. Miller

Chart times (local, standard):
- 12:00 mid. May 6
- 11:00 pm May 21
- 10:00 pm June 6
- 9:00 pm June 21

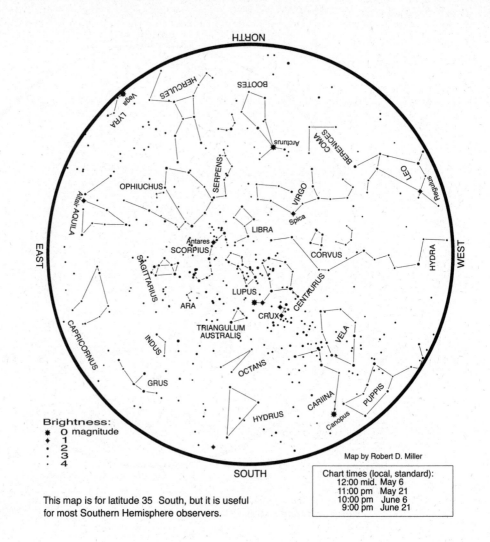

NORTH

EAST

WEST

SOUTH

LYRA
Vega
HERCULES
BOOTES
Arcturus
COMA BERENICES
LEO
Regulus
OPHIUCHUS
SERPENS
VIRGO
Spica
Altair AQUILA
LIBRA
HYDRA
Antares
SCORPIUS
CORVUS
SAGITTARIUS
LUPUS
CENTAURUS
ARA
CRUX
CAPRICORNUS
TRIANGULUM AUSTRALIS
VELA
INDUS
OCTANS
PUPPIS
GRUS
CARIINA
CANOPUS
HYDRUS

Brightness:
* 0 magnitude
1
2
3
4

Map by Robert D. Miller

Chart times (local, standard):
12:00 mid. May 6
11:00 pm May 21
10:00 pm June 6
9:00 pm June 21

This map is for latitude 35 South, but it is useful
for most Southern Hemisphere observers.

Appendix B

Glossary

antimatter: Matter composed of antiparticles that have the same mass but opposite electrical charge of ordinary particles.

asterism: A named pattern of stars, such as the Big Dipper, that isn't one of the 88 official constellations.

asteroid: One of many small, rocky, and/or metallic bodies orbiting the Sun.

aurora: The occurrence of a colorful display of light in the upper atmosphere of Earth or another planet, produced by the collision of electrically charged particles with gas atoms and molecules.

binary star: Two stars orbiting around a common center of mass in space; also called a binary system.

black hole: An object with a gravity so strong that nothing inside can escape — not even a ray of light.

bolide: A very bright meteor that appears to explode or that produces a loud noise.

comet: One of many small bodies made of icy and dusty matter orbiting the Sun.

constellation: Any one of 88 regions on the sky, typically named after an animal, object, or ancient deity (for example, Ursa Major, the Great Bear). All stars within the boundaries of the region are part of the constellation.

cosmic rays: High-energy subatomic particles from the Sun (solar cosmic rays), the Milky Way (galactic cosmic rays), or even beyond the Milky Way (extragalactic cosmic rays).

crater: A round depression on the surface of a planet, moon, or asteroid created by the impact of a falling body, a volcanic eruption, or the collapse of an area.

dark energy: An unexplained physical process that acts as if it were a repulsive force, causing the universe to expand at a greater and greater rate as time goes on.

dark matter: One or more unknown substances in space that exert gravitational force on celestial objects, which is how astronomers detect its existence.

Doppler effect: The process by which light or sound is altered in perceived frequency or wavelength by the motion of its source with respect to the observer.

double star: Two stars that appear very close to each other on the sky and that may be physically associated (a binary star) or may be unrelated to each other and at different distances from Earth.

dwarf planet: An object that orbits the Sun, isn't the moon of a planet, is massive enough that its own gravity makes it round, and hasn't cleared its orbit of other small bodies. Pluto is a dwarf planet.

eclipse: The partial disappearance (partial eclipse) or total disappearance (total eclipse) of a celestial body when another object passes in front of it or when it moves into the shadow of another object.

ecliptic: The apparent path of the Sun across the background of the constellations.

exoplanet: A planet of a star other than the Sun. Also called an *extrasolar planet.*

fireball: A very bright meteor.

galaxy: A huge system of millions or billions of stars, sometimes with vast amounts of gas and dust, enveloped in a large region (halo) of dark matter.

gamma ray burst: An intense outburst of gamma rays that comes without warning from a random spot in the distant universe.

meteor: The flash of light caused by the fall of a meteoroid through Earth's atmosphere; the term *meteor* is often incorrectly used to refer to the meteoroid itself.

meteorite: A meteoroid that landed on Earth.

meteoroid: A rock in space, composed of stone and/or metal; probably a chip from an asteroid.

near-Earth object: An asteroid or comet that follows an orbit that brings it close to Earth's orbit around the Sun.

nebula: A cloud of gas and dust in space that may emit, reflect, and/or absorb light.

neutrino: A subatomic particle that has no electric charge and an extremely small mass. It can pass through a whole planet or even the Sun.

neutron star: An object only tens of miles across but greater in mass than the Sun (all pulsars are neutron stars, but not all neutron stars are pulsars). The dead, hot core of a star more massive than the Sun.

OB association: A loosely clustered group of young, hot stars.

occultation: The process by which one celestial body passes in front of another, blocking it from the view of an observer.

orbit: The path followed by a celestial body or a spacecraft.

planet: A large, round object that forms in a flattened cloud around a star and — unlike a star — doesn't generate energy by nuclear reactions.

planetary nebula: A glowing, expanding gas cloud that was expelled in the death throes of a Sun-like star.

pulsar: A fast-spinning, tiny, and immensely dense object that emits light, radio waves, and/or X-rays in one or more beams like the beam from a lighthouse.

quasar: A small, extremely bright object at the center of a distant galaxy, thought to represent the emission of much energy from the surroundings of a giant black hole.

red giant: A large, very bright star with a low surface temperature; also a late stage in the life of a Sun-like star.

redshift: An increase in the wavelength of light or sound, often due to the Doppler effect or, in the case of distant galaxies, the expansion of the universe.

rotation: The spinning of an object around an axis that passes through it.

seeing: A measure of the steadiness of the air at a place of astronomical observation (when the seeing is good, the images you view through telescopes are sharper).

SETI: The Search for Extraterrestrial Intelligence, a program of radio astronomy observations (and other observations) that seeks to detect messages from intelligent civilizations elsewhere in space.

solar activity: Changes in the appearance of (and in the radiation from) the Sun that occur from second to second, minute to minute, hour to hour, and even year to year. It includes eruptions such as solar flares and coronal mass ejections and other features such as sunspots.

spectral type: A classification applied to a star based on the appearance of its spectrum, usually related to the temperature in the region where the visible light from the star originates.

star: A large mass of hot gas held together by its own gravity and fueled by nuclear reactions.

star cluster: A group of stars held together by their mutual gravitational attraction that formed together at about the same time (types include globular clusters and open clusters).

sunspot: A relatively cool and dark region on the visible surface of the Sun, caused by magnetism.

supernova: An immense explosion that disrupts an entire star and that may form a black hole or a neutron star.

terminator: The line separating the illuminated and dark parts of a celestial body that shines by reflected light; the boundary between day and night.

transit: The movement of a smaller object, such as Mercury, in front of a larger object, such as the Sun.

variable star: A star that changes perceptibly in brightness.

white dwarf: A small, dense object shining from stored heat and thus fading away; the final stage in the life of a Sun-like star.

zenith: The point on the sky that's directly above the observer.

Sky Measures

arc minutes/arc seconds: Units of measurement on the sky. A full circle around the sky consists of 360°, each divided into 60 arc minutes; each arc minute is divided into 60 arc seconds.

astronomical unit (AU): A measure of distance in space, equal to the average distance between Earth and the Sun — about 93 million miles.

declination: On the sky, the coordinate that corresponds to latitude on Earth and that's measured in degrees north or south of the celestial equator.

light-year: The distance light travels in a vacuum (for example, through space) in one year; about 5.9 trillion miles.

magnitude: A measure of the relative brightness of stars, with smaller magnitudes corresponding to brighter stars. For example, a 1st magnitude star is 100 times brighter than a 6th magnitude star.

right ascension: On the sky, a coordinate that corresponds to longitude on Earth and that's measured eastward from the vernal equinox (a point on the sky where the celestial equator crosses the ecliptic and where the Sun is located on the first day of spring in the Northern Hemisphere).

Index

A

absolute magnitude, 55
accretion disk, 218, 271–272
acidophiles, 305
active galactic nuclei (AGN), 276–281
age of universe, 330–332
Ahnighito meteorite, 142
albedo maps, 136
Algol (Beta Persei), 236
aliens. *See* SETI
alkalophiles, 305
Allen Telescope Array (ATA), 291
Alpha Canis Majoris (Sirius), 238
Alpha Centauri C (Proxima Centauri), 20, 212, 233, 237–238, 300–302
Alpha Centauri system, 237–238
Alpha Lyrae (Vega), 238
Alpha Magnetic Spectrometer (AMS-02), 316, 317, 320
Alpha Orionis (Betelgeuse), 238
alt-azimuth mounts, 62, 63
amateur astronomers, 1, 9, 43. *See also* citizen science projects
American Association of Variable Star Observers (AAVSO), 239
American Museum of Natural History, 38, 142
ancient astronomy, 9–13
Andromeda Galaxy, 21, 245, 260
angular diameter, of Venus, 133
anisotropy, 328
annihilation radiation, 320
annular solar eclipses, 202
antimatter, 309, 319–320, 353

antiprotons, 320
aperture, 57, 60, 64, 199
apogee, of Moon, 109
apparent magnitude, 53, 55
apparition, of planets, 130
apps, 18, 33–34, 50
arc minutes, 22, 23, 133, 356
arc seconds, 22, 23, 24, 133, 356
artificial satellites, 88–91
ashen light, Venus, 134
associations, in galaxies, 243
asterisms, 8–9, 353
asteroid belt, 141–145, 341
asteroidal meteoroids, 70, 142
asteroids
  common errors about, 341
  defined, 70, 141, 353
  naming, 88
  nudging, 147–148
  observing, 149–151
  occultation, 112, 150–151
  striking Jupiter, 160
  threat of, 145–149
astrobiology, 304–307
AstroFests, 40
astronomical unit (AU), 21, 356
astronomy
  ancient, 9–13
  common errors about, 339–342
  defined, 1, 8
  getting into gradually, 66–68
  observational nature of, 8–9
  overview, 1–4, 7–8
astronomy apps, 18, 33–34, 50
astronomy clubs, 30–31, 67
astronomy magazines, 32

Astronomy on Tap events, 40
astronomy software, 33–34
astronomy vacations
  AstroFests, 40
  Astronomy on Tap events, 40
  dark sky parks, 44–46
  eclipse cruises and tours, 40–42
  overview, 38
  star parties, 39–40
  telescope motels, 42–44
astronomy websites, 31–32
Astropulse project, 292
aurorae, 97–98, 191, 353
Automated Planet Finder (APF), 291, 295
averted vision, 133
axions, 315
axis, of Earth, 104

B

Backyard Worlds: Planet 9 project, 179
Baily's Beads, 204–205
bandwidth, H-alpha filters, 198
barophiles, 305
barred spiral galaxies, 243, 257–258
baryonic dark matter, 314
Beehive (Praesepe) cluster, 247
Beta Persei (Algol), 236
Betelgeuse (Alpha Orionis), 238
Big Bang
  age of universe, 330–332
  common errors about, 342
  cosmic microwave background, 322–323, 327–329

Earth *(continued)*

space weather, 191–192

tilt of, 103–105

time on, 102–103

unique characteristics of, 96–97, 336

earthshine, 110, 115–116

eclipse cruises and tours, 40–42

eclipses

defined, 354

of Jupiter moons, 159

lunar, 109, 110–111

solar, 40–42, 202–206

stellar, 235–236

eclipsing binary stars, 235–236

ecliptic, 49, 105, 131, 354

Eight-Burst Nebula, 256

Einstein, Albert, 25–26

Einstein ring, 312–313

elliptical galaxies, 257, 258–259

elongation, of planets, 130–132

Enceladus satellite, 165, 307

end states of stellar evolution

black holes, 217–219

central stars of planetary nebulae, 214

neutron stars, 216–217

overview, 210, 213

supernovas, 215–216

white dwarfs, 214–215

ephemeris, 25

equatorial mounts, 62, 63

equinoxes, 105

Eros asteroid, 145

escape velocity, 268

Eta Carinae, 338

Europa moon, 157–160, 306

European Space Agency, 81, 124. *See also specific missions or projects*

evening star, 129, 130

event horizon, 219, 269–271, 275

evolution of universe. *See* Big Bang

exo-Earths, 298

exoplanets

changing ideas on, 294–295

checking out, 303–304

defined, 354

finding, 295–298

orbiting Proxima Centauri, 300–302

orbiting TRAPPIST-1, 302

overview, 293

as targets for SETI, 292

types of, 298–300

expansion of universe, 27, 322, 330–332

exploding stars, 233–235

extraterrestrial beings. *See* SETI

extremophiles, 304–305

extrinsic variable stars, 230

F

falling objects, black holes, 269, 271

FAST (Five-hundred-meter Aperture Spherical radio Telescope), 292

fast telescopes, 86

filaments, on Sun, 198

filters, solar, 64, 197, 198–199, 203

finder charts, 85

fireballs, 72–73, 77, 354

first contact, 202

flare stars, 232–233

focal length, telescope, 62

focal point, binoculars, 58–59

Ford, Kent, 310–311

47 Tucanae cluster, 249–250

fossil evidence, on Mars, 127–128

Foucault pendulums, 49

fourth contact, 205

front-end filters, 198–199

full-aperture solar filters, 199

fusion, nuclear, 184

G

Gaia satellite, 245

galactic bulge, 243, 281, 311

galactic center, 243

galactic disk, 243, 244

galactic equator, 244

Galactic Latitude, 244

Galactic Longitude, 244

galactic orbiting, 26, 27

galactic plane, 244, 245

galactic rim, 245

galactic year, 26

galaxies. *See also specific galaxies; specific galaxy types*

clusters of, 259, 263–264, 312, 318–319

cold dark matter, 313–314

cosmic voids, 264–265

dark matter in, 310–312

defined, 354

fate of universe, 332

Great Walls, 264–265

Hubble constant, 331

Hubble's Law, 331

Local Group, 263

nebulae, 250–256

overview, 241

quasars in, 278

star clusters, 246–250

superclusters, 264–265

types of, 256–260

viewing, 260–265

visible with naked eye, 245

weak lensing, 313

Galaxy Zoo, 34, 265

Galilean moons, 157–160

Galilei, Galileo, 194, 200

gamma ray burst, 354

New Horizons probe, 171, 175, 176

Newton, Sir Isaac, 25

Newtonian reflectors, 60, 194–197

NEXT satellites, 90

NGC (New General Catalogue) number, 18

NGC 205 galaxy, 260–261

NGC 4755 (Jewel Box) cluster, 248

NGC 6231 cluster, 248

North American Nebula, 255

North Celestial Pole (NCP), 22, 24, 48

north magnetic polarity, on Sun, 190

North Star (Polaris), 48, 50–52, 104, 230

Northern Coal Sack nebula, 255

Northern Cross asterism, 234

northern lights. *See* aurorae

novas, 233

nuclear fusion, 184

nucleus, comet, 80–81

O

OB associations, 250, 354

objective lenses, binoculars, 56–57

observable universe, 209

observatories, 34–37

observing meetings, 30

occultations, 111–112, 150–151, 159, 354

ocean, on Mars, 124–125

ocean tides, on Earth, 336

oddball dark matter, 315

off-axis solar filters, 199

Olympus Mons, Mars, 126

Omega Centauri cluster, 249–250

Oort Cloud, 79–80

open clusters, 246–248

opposition, of planets, 130–132, 135

optical doubles, 228

Optical Gravitational Lensing Experiment (OGLE), 296

optical magnification, 57–58

optically violently variable quasars (OVVs), 279

orbit, defined, 354

orbital plane, 235

orbital velocity, 226–228, 235–236

Orion constellation, 52

Orion Nebula, 211, 254

oxygen, on Earth, 96

P

Palomar Observatory, 36

Panoramic Survey Telescope and Rapid Response System (Pan-STARRS), 149

parent atoms, 106

Parkes Radio Telescope, 37

parsecs, 263

partial lunar eclipses, 111

partial solar eclipses, 202

path of totality, 41, 111, 205–206

penumbra, 200, 202, 203

Penzias, Arno, 323

perihelion points, 178

period-luminosity relation, 231

Perseids meteor shower, 74

photinos, 315

photographing meteors, 78–79

photosphere, Sun, 185–186

pinhole camera, 197

Pinwheel (Triangulum) Galaxy, 245, 261

Planck satellite, 329

Planet Hunters project, 304

Planet Nine, search for, 178–179

planetarium programs, 33

planetariums, 34, 38

planetary nebulae, 193, 214, 252–253, 354

Planetary Photojournal site, 123, 157, 177

planetary systems, 293

planetary transit, 135

planets. *See also specific planet types; specific planets*

comparative planetology, 101, 128–129

defined, 354

good seeing, 66

naked-eye observation, 53

overview, 95

search for ninth, 178–179

sky geography, 49–50

versus stars, 10

planispheres, 50

plasma, lack on Earth, 336

plasma tail, comets, 82–83

plate tectonics, 96

Pleiades (Seven Sisters) cluster, 247

Plutinos, 175

Pluto, 170–175, 178, 337

Polaris (North Star), 48, 50–52, 104, 230

polarities, on Sun, 190

polarized radio waves, 279

Ponticus, Heraclides, 48

porro prisms, 57

positrons, 320

potentially hazardous asteroids (PHAs), 145

Praesepe (Beehive) cluster, 247

prisms, binoculars, 56–57

prograde orbits, 164

Project Ozma, 286–287

Project Phoenix, 290–291

projection technique, 194–197

prominences, on Sun, 187, 188–189, 198

protoplanetary nebulae, 253

Proxima b, 300–302

water worlds, 299

weak lensing, 313

websites, astronomy, 31–32

Wesley, Anthony, 160

West comet, 85

Whirlpool Galaxy, 261, 262

white dwarfs, 193–194, 214–215, 223, 224, 355

white light solar filters, 197, 198–199

white-light photographs, 206–207

Wild-2 Comet, 71, 81

Wilkinson Microwave Anisotropy Probe (WMAP), 329

Wilson, Robert, 323

WIMPs (weakly interacting massive particles), 315, 316–317

wobble method, for finding exoplanets, 295

wormholes, 271

wrongway planets. *See* retrograde orbits

X

X-ray telescopes, 272

Y

young stellar objects (YSOs), 210, 211–212

Z

Zeeman effect, 229

zenith, defined, 355

Zodiac constellations, 49

Zwicky, Fritz, 310, 311

About the Author

Stephen P. Maran, PhD, a 36-year veteran of the space program, has been honored with the NASA Medal for Exceptional Achievement, the Klumpke-Roberts Award of the Astronomical Society of the Pacific for "outstanding contributions to the public understanding and appreciation of astronomy," the George Van Biesbroeck Prize of the American Astronomical Society for "long-term extraordinary or unselfish service to astronomy," and the Andrew W. Gemant Award of the American Institute of Physics for "significant contributions to the cultural, artistic, or humanistic dimension of physics."

In 2000, the International Astronomical Union named an asteroid (Minor Planet 9768) "Stephenmaran" for Dr. Maran.

Dr. Maran began practicing astronomy from rooftops in Brooklyn and at a golf course in the Bronx. He graduated to conduct astrophysical studies at Kitt Peak National Observatory, the National Radio Astronomy Observatory, and Palomar Observatory in the USA, and Cerro-Tololo Inter-American Observatory in Chile. He also conducted research with instruments in space, including the Hubble Space Telescope (HST) and the International Ultraviolet Explorer. He helped design and develop two instruments that flew in space aboard the HST. He also taught astronomy at the University of California, Los Angeles and at the University of Maryland, College Park.

As press officer of the American Astronomical Society for many years, Maran presided over media briefings that brought news of astronomical discoveries to people worldwide. He also chaired televised briefings at NASA and, in 2015, came out of retirement to work as a science writer embedded with the New Horizons science team during the historic encounter of NASA's New Horizons probe with Pluto and its moons.

Dr. Maran observed total eclipses of the sun from the Gaspé Peninsula and elsewhere in Quebec; Baja California in Mexico; the Sahara desert in Libya; at sea off New Caledonia and Singapore in the eastern Pacific; and in the United States.

In the course of spreading the good word about astronomy, Maran has lectured on black holes in a bar in Tahiti and explained an eclipse of the sun on NBC's *Today* show. He's served as an astronomy lecturer aboard ships of the Celebrity, Cunard, MSC Cruises, and Sitmar lines, including cruises to view solar eclipses and Halley's comet. He's addressed audiences ranging from inner-city Seattle school children and Atherton, California, Girl Scouts to the National Academy of Engineering in Washington, DC, and subcommittees of both the U.S. House of Representatives and the U.N. Committee on the Peaceful Uses of Outer Space.

Dr. Maran is an author or editor of ten other books on astronomy; has written many articles on astronomy and space exploration for *Natural History*, *Smithsonian*, and other magazines; and served as a writer and consultant for the National Geographic Society and Time-Life Books. He's currently a consultant for Harvard University Press.

Dr. Maran graduated from Stuyvesant High School in New York City and Brooklyn College. He received both his MA and PhD in astronomy from the University of Michigan. He's married to Sally Scott Maran, a journalist. They have three children and four grandchildren.

Dedication

To Sally, Michael, Enid, and Elissa with all my love.

Author's Acknowledgments

Thanks first to my family and friends who put up with me in the writing of this book and its new edition. Thanks also to my agent, Skip Barker, who has aided me from the inception of the project. I thank Stacy Collins for her faith in the original project and Lindsay Lefevere for championing both the third edition of *Astronomy For Dummies* and this substantially revised and updated fourth edition.

I'm grateful to Ron Cowen and Dr. Seth Shostak for their past contributions to this book; to Vicki Adang and Megan Knoll, who organized and edited it; and to their skilled colleagues on the editorial and production teams at Wiley Publishing who make each new edition better and brighter. I'm especially grateful to professor Misty Bentz, who carefully read the manuscript and pointed out dozens of places where improvements were in order. Any remaining error is mine alone.

Thanks also to the organizations that provided the photographs in this book and to the producer of the star maps, Robert Miller.

I'm grateful to several experts for prompt answers to my queries about topics in this new edition. In particular I thank Drs. Joseph Gurman, Marc Kuchner, and John Mather (all of NASA Goddard Space Flight Center); Prof. Lynn Cominsky (Sonoma State University); Drs. Richard Fienberg and Kevin Marvel (both of the American Astronomical Society); Messrs. David Finley (National Radio Astronomy Observatory) and Mike Hankey (American Meteor Society); Dr. Scott McIntosh (High Altitude Observatory); Dr. Harold Weaver (Johns Hopkins University Applied Physics Laboratory); and Dr. Donald Yeomans (Jet Propulsion Laboratory).

Publisher's Acknowledgments

Executive Editor: Lindsay Sandman Lefevere
Development Editor: Victoria M. Adang
Copy Editor: Megan Knoll

Technical Editor: Dr. Misty C. Bentz
Production Editor: Siddique Shaik
Cover Photo: © Allexxandar/Getty Images